Kissed by a Fox

Kissed by a Fox

AND OTHER STORIES OF
FRIENDSHIP IN NATURE

Priscilla Stuckey

COUNTERPOINT PRESS | BERKELEY

Library of Congress Cataloging-in-Publication Data is available

ISBN: 978-1-58243-812-2

Cover Design by Ann Weinstock
Interior design by meganjonesdesign.com

COUNTERPOINT
1919 Fifth Street
Berkeley, CA 94710
www.counterpointpress.com

Distributed by Publishers Group West

10 9 8 7 6 5 4 3

For the Earth

—every one of us

CONTENTS

The speech of water, the speech of earth, and the speech of mud
Are heard by those who listen with the heart.

—RUMI, MASNAVI

⌒

We are most ourselves when we are most intimate with the rivers
and mountains and woodlands, with the sun and the moon and the
stars in the heavens; when we are most intimate with the air we
breathe, the Earth that supports us, the soil that grows our food,
with the meadows in bloom. We belong here. Our home is here.

—THOMAS BERRY, THE SACRED UNIVERSE

⌒

When our beliefs settle down to sleep and the streetlights come
on, if we said matter was holy, would we then love and be joyous?

—LINDA HOGAN,
THE WOMAN WHO WATCHES OVER THE WORLD

PROLOGUE

Bald Eagle

MIDNIGHT BLUE WATER rested at the horizon under a brightening sky, framing shafts of pine and fir at shore's edge. Here on the southern tip of Lopez Island, off the coast of Seattle, the dawn air was cool and still, the only sound a few songbirds calling far away. I headed across the needle-packed yard toward a clump of pines. The bald eagle nest was right over there, Anya had said; she'd seen the chicks fledging just days ago, and they couldn't be far away now.

I turned my binoculars toward the pines, eager to spot that huge platform of sticks three to five feet across. Eagles use the same structure year after year, weaving in more and more sticks for support until the whole can weigh a ton or more. Surely a nest the size of a mattress should be easy to

1

find! I scanned the trees in one direction. Nothing. Puzzled, I looked the other way. No nest anywhere in sight.

Soon my neck stiffened from the upward gaze. I lowered the glasses and headed across the thick carpet of pine needles. The eagles would have to show up soon. Anya had said they were here, so I might as well wait.

It was 1995, and I'd met Anya just a few days earlier on a women's camping trip to Mount Rainier, a vacation from my too-quiet life in Oakland, California, where I lived alone writing a doctoral dissertation and editing books. A few years earlier, at thirty-five, I'd undergone three severe losses—deaths of both parents plus divorce after thirteen years of marriage—followed too quickly by the end of a new relationship, and now I spent most days treading the waters of depression, withdrawn and silent. On the camping trip Anya had been almost as quiet. Each morning she would unroll her yoga mat a few feet away from the group and stretch silently through her poses. We'd hardly said a word to each other. Yet on the last day, when the group gathered on soft emerald grass next to a tiny ripple of creek to share what we'd gained from the trip, and I said I'd loved every minute of it but still—in this prime bald eagle country—hadn't seen an eagle, Anya had urged me to follow her home. Her guest room was really a workout room, but I was welcome to unroll my sleeping bag there for a couple of nights and search for eagles by day.

Climbing now over rocks near the water's edge, I gazed toward the deep blue of the sea. No eagles. I wondered how to find them. The thought occurred: *Why don't you call them?* While meditating recently, I'd seen trees and birds in my mind's eye, like watchful presences, but that was meditation, not "real" life. Birds don't just come when you call. True, my skeptical biologist friend Meredith, who had taught me birding, had told me how, during a bewildering time, she'd gone to the beach and asked for an osprey to appear. Although it was not the season for osprey, and they were fairly rare at that beach, within minutes an osprey had soared high overhead. Nice coincidence, I'd thought, but no rational evidence suggested that birds could hear or respond.

Still, I was going to be here only until tomorrow, so there was no time to waste.

I sat down on a soft bed of pine needles, closed my eyes, and took a deep breath. Inside, thoughts ran wild: *This is silly . . . They'll never show up . . . Might work for other people, not you.* I took another deep breath and settled into my body again, feeling alive in my arms, in my legs and feet. Noticing sensations was feeling, not thinking—a good way to turn off mental chatter.

I brought to mind a picture of a bald eagle: that white head, fierce yellow eyes, and hooked beak; the dark body, white tail, and powerful talons that could make off with a fish nearly half the bird's size. *Please,* I spoke mentally to the image, *I'd really like to see you. Is it possible for you to hear me?*

In the background I could hear another voice: *Yeah, sure, right—now you're praying to birds? Talking to someone you can't see, who may not even be there?*

I breathed again. *I'm only going to be here today and part of tomorrow. I've seen lots of birds but never bald eagles. Please let me catch a glimpse of you.* I remained quiet for a few moments, eyes closed.

There was no more to do. I opened my eyes. The sun was far above the horizon now. I sat in its midmorning warmth gazing quietly at pines. How long do you wait for an eagle to come when you call?

If you're impatient like me, not very long. A minute ticked by, then two, then five. Nothing happened.

Trying not to notice a corner of disappointment, I stood up and headed toward the bike shop across the way. My chances of seeing eagles would increase—wouldn't they?—if I covered the island. I picked out a bicycle and at the café next door stocked up with a sandwich, water, and an apple.

Map in hand, new Lopez Island visor peeking out from under my bike helmet, I headed out.

The road I chose took me inland, and soon I was engulfed by peaceful summer-gold fields stretching to the horizon, the sea no longer visible beyond their gentle undulations. Cars passed at a leisurely rate, each driver in turn lifting an index finger—not four fingers, not a palm, just an index finger—to greet me. Shyly I waved back, unaccustomed after a decade of urban life to waving at strangers. I might have been back in the rural Ohio of my childhood.

I covered the length of the island, stopping for lunch along a sand spit stretching across a still lagoon, then heading inland again for more miles of quiet fields. Now and then I scanned the horizon for the telltale sign of black wings spread wider than a hawk's and flattened out horizontally from the body, not lifted in a V like a vulture's. I saw golden grain and craggy cliffs with pine trees silhouetted black against the blue sound, but I saw no eagles' wings.

By late afternoon my bottom was sore from the thick denim seam in my jeans—I hadn't packed biking shorts—and my leg muscles were yelping. Drinking the last of my water, I worried that I wouldn't get back to my sleeping bag before dropping from exhaustion.

Finally back at Anya's house, I finished the other half of my sandwich, soggy from the backpack. At dusk Anya returned, as quiet as she'd been on the camping trip. I told her no luck, I hadn't seen any eagles. She said, Maybe tomorrow.

I went to bed wondering what had gone wrong. *I guess beginners don't get what they ask for*, I thought. The next morning I would have time only to turn in my bike at the shop and head to the ferry landing for the return trip to Seattle and the flight home. The corner of disappointment had grown throughout the day and now threatened to take over.

Up early again the next morning, I thanked Anya for the impromptu visit and took one last hike around the outside of the house. Binoculars in hand, I searched again for the nest. If an eagle nest is so obvious, why couldn't I spot it?

I called again to the eagles. *And please,* I added, hoping I didn't sound too demanding, *it has to be soon, because I have to leave right now!*

At the bike shop a guy in a baseball hat pointed to the best corner for hitching a ride across the island, so I headed there and stuck out my thumb. Anya had assured me that Lopez Islanders often drove tourists to the ferry landing.

One car went by, then two. I tightened the straps on my pack and prepared for the three miles ahead. I was well out of sight of the hamlet, enjoying again the narrow road stretched between green and gold fields, before the next car approached. I turned and stuck out my thumb, and the small convertible pulled to a stop beside me. The woman driving waved me in cheerfully, and I hefted my pack over the side and settled gratefully down. Sun warmed the backs of our necks as we made our way north.

We were now in the very middle of the island. No other cars passed us on the road, there were no trees nearby, and we were sailing past tranquil fields, our vision open to the sky. I glanced to the right and saw, far away, across the fields, above the tree line at the eastern horizon, a tiny black spot in the sky. I ripped my binoculars from my pack and focused. Those spread-wide wings! Closer it drew. A nearly invisible tail! A light-colored head!

Over here! My silent call was more wish than thought. A head and tail of white materialized in the lenses. The bald eagle was headed straight toward us.

"Could you stop for a minute?" I asked the driver. She too had spotted the faraway shape to our right. "Sure," she said, smiling, as she braked to a halt in the middle of the road. Still the eagle approached, on a course that would take it directly over our heads.

What happened next made my heart stop—and does so even now, more than fifteen years later. As we watched, the bald eagle, instead of flying over us and across the island, spotted the car below and, when it had arrived directly overhead, turned sharply to make a tight circle. I raised the binoculars again, eager for a closer look. There were the broad dark wings, spread white tail, pure white head. There were the sharp eyes trained on us as the eagle turned in orbit, shifting its regal head to stare directly down into the car. I watched, first with the glasses, then without. I had plenty of time to take in every detail, for the eagle, instead of making only one circle, was turning in many tight circles directly overhead. *Thank you!* I whispered. *Thank you, oh, thank you!* My heart was rising in my chest, as if elation could lift it to the eagle's height.

I watched until I couldn't hold my head up anymore. Only then did the eagle break orbit, beat its wings three or four times, and head back toward the spot on the horizon from which it had appeared.

↝

STILL IN A daze, I boarded the Anacortes ferry and rode across the strait. I stared at the blue water, unable to think of anything but the miracle I had just witnessed. With my mind lost in wonder, my senses were freed to feel the sharp bite of sea breeze, to hear the piercing calls of gulls, to see in the deep water of midday the deep blue of nightfall.

On the two-hour bus ride to the airport, my mind dwelled still on the eagle making its tight circles above our stopped car. When I got home, there would still be a dissertation to write, friends would still be scarce and finances tight, but this one thing I knew: someone had heard my call. Something new was possible. The despair I had been fitted with like a suit of family clothes had been torn by a visit from another species. Although it would be years before the despair fell away completely, receiving a visit from the eagle nourished my growing sense of wonder. There was magic in the world, and it could be called up by sincere wishes sent out with respect. The eagle had answered—at a time when few others in my life were responding. Perhaps the world was a friendly place after all.

~

THAT A WILD animal had come to my call was a smack to my consciousness. Something was afoot in the world—something that the best descriptions of that world, thought up by the most brilliant minds of the most modern society in that world, could not explain and did not even have words for.

Once I was home again, to say I turned skeptical of what I had experienced is to put it mildly. After all, I was thirty-eight years old and a doctoral candidate at the Graduate Theological Union in Berkeley. Through a decade of graduate school I'd been trained as a meticulous observer and in my dissertation was exploring gender and nature in Western religions. My school valued both rationality and faith, and it promoted faiths of all kinds, from Christianity to Judaism to Buddhism. Yet nothing in my graduate studies talked about what I had just experienced. Not one of those great religions taught that humans could become intimate with other animals or plants. There was no language for my experience, and so for all practical purposes my experience didn't exist. Interspecies communication was not "spirituality"; it belonged instead to the back alleys of parapsychology and New Age woo-woos, which means—this is an understatement—that it was not taken seriously. Certainly I never mentioned the eagle to other students, much less my professors. After all, I hardly believed myself what had happened.

Yet by that point I'd also been a feminist for more than fifteen years, and if feminism had taught me anything, it was to pay attention to my own experience—to trust it even when signals from others contradict it, to hold my own way of seeing as equal in value to the seeing of others. Feminism, in other words, had taught me courage. Little in Christianity, with its doctrine of original sin—that fatal flaw in humanity that means you can never fully trust yourself—supported or even recognized this form of courage; Christian feminists had been saying so for decades already. My favorite Zen axiom had become, "If you meet the Buddha in the road, kill him," which is said to mean, Don't believe what even the most revered authorities tell you, but instead test everything—*everything!*—in your own experience.

So when my experience started diverging from mainstream rationalism, including the religious kinds, I could not completely discount it. I had to pay attention. The eagle had bestowed a tremendous blessing, and being faithful to my experience meant being faithful also to this being's gift. The eagle had flown across the island in response to my need, and fidelity to that appearing meant allowing the eagle to make a difference in my life; it meant, in effect, opening myself to a new kind of faith.

ç

WE DON'T USUALLY think of our worldview as our faith; "having faith" more often means belonging to a religious group or believing in God or even feeling certain that something fantastic is about to happen; Red Sox fans in 2004 were paragons of faith.

But another meaning of *faith* is our basic stance in life—the worldview that orients us. It's like a travel guide written by our culture and handed to us at birth to help us make sense of our adventures along the way. Like any guidebook, it points us toward some paths and glosses over others; it says "stop here" but "don't bother over there." And in 1995, when I was a doctoral student, my guidebook said clearly that animals do not hear human thoughts. It also hinted that anyone who disagrees is probably just a teensy bit crazy. The authors of this guidebook, religious and scientific minds alike, agreed that humans are different from the rest of the creatures; we alone possess mind—at least the kind of mind that can solve problems or make tools or feel emotions. And even if scientific research showed those criteria crumbling and falling away, at least we could take refuge in our human-only use of language. To put it simply, plants and animals, let alone rocks and earth, don't talk. And they surely don't pay attention to what humans are thinking and feeling. On this score my guidebook was clear.

That the rest of the creatures are mute is, I have come to see, an article of faith, not a conclusion about the world reached through examining the evidence. It's a relic of seventeenth-century Europe, where people became gripped with the idea that nature works like a machine, and it grew into the Western consensus that mind and matter are separate. But when I was thirty-eight and a graduate student, my faith in this consensus had begun to shift, and the bald eagle's appearance was but one of the hard-to-explain events—"fissures in ordinary logic," writer Carol Flinders calls them—that was compelling me to seek a different guidebook.

Philosopher of science Thomas Kuhn might have called it an anomaly—an event that cannot be fitted into the system of shared beliefs called a paradigm. Kuhn is famous for pointing out that paradigms are chosen, not proved. There is no irrefutable evidence to support one paradigm over another; it "can never be unequivocally settled by logic and experiment alone." This does not mean the choice is irrational, only that it depends on factors, and especially values, lying outside the bounds of science.

Some distance always remains between a paradigm and the "real world." Paradigms are not nature itself but rather humanly created models of nature. Travel guides, not the territory.

⌒

IF KUHN WAS right, then the door is open to reconsidering our paradigm of nature. The view of nature as a machine, lacking mind or spirit, can be seen as a choice, the product not of logic and evidence but of values that lie outside science itself. In fact, this is what historians find when they look into the lives of the seventeenth-century thinkers who set this worldview in place, such as Francis Bacon and René Descartes. Descartes, for instance, instituted the laboratory practice of vivisection—dissecting an animal such as a dog while alive and conscious—not because he had evidence that animals feel no pain but because he *began* with the belief that animals lack soul or consciousness and *therefore* cannot feel pain. He carried his beliefs into the laboratory rather than deriving his beliefs from what he observed.

Moreover, our ruling paradigm of matter as devoid of mind or spirit may not at all describe "nature as it really is." While being tremendously useful for solving certain puzzles, it leaves others completely unexplained. In the laboratory, researchers find subatomic particles behaving as though they could have mental properties, or mice showing empathy for cagemates in pain—anomalies that stretch the contours of what we normally expect. And in everyday life, an eagle showing up in response to a telepathic call simply doesn't fit what we think we know about the world.

In the twenty-first century, we are beginning to see that relying on a four-hundred-year-old paradigm has led us also to the brink—perhaps beyond—of ecological disaster. If mind belongs to humans alone, then stones, trees, and streams become mere objects of human tinkering. We can plunder the Earth's resources with impunity, treating creeks and mountaintops in Kentucky or rivers in India or forests in northwest North America as if they existed only for economic development. Systems of land and river become inert chunks of lifeless mud or mechanical runs of H_2O rather than the living, breathing bodies upon which we and all other creatures depend for our very life.

Not to mention what "nature as machine" has done to our emotional and spiritual well-being. When we regard nature as churning its way forward mindlessly through time, we turn our backs on mystery, shunning the complexity as well as delights of relationship. We isolate ourselves from the rest of the creatures with whom we share this world. We imagine ourselves the apex of creation—a lonely spot indeed. Human minds become the measure of creation and human thoughts the only ones that count. The result is a concept of mind shorn of its wild connections, in which feelings become irrelevant, daydreams are mere distractions, and nighttime dreams—if we attend to them at all—are but the cast-offs of yesterday's overactive brain. Mind is cut off from matter, untouched by exigencies of mud or leaf, as if the human mind were not, like trees, shaped by whispers or gales of wind, as if we were not, like rocks, made of soil.

And then we wonder at our sadness and depression, not realizing that our own view of reality has sunk us into an unbearable solipsism, an agony of separateness—from loved ones, from other creatures, from rich but unruly emotions, in short, from our ability to connect, through sense and feeling and imagination, with the world that is our home. We stand in self-decreed exile, having lopped off our awareness of the mind flowing through all creatures in this world, and so we are unable to know ourselves as kin with those creatures, participants in the same life permeating all.

MANY SOCIETIES, BOTH past and present, choose a different worldview: they count plants, animals, or mountains as friends. It is a view that Western scholars call "animist." But *animist* is so abstract a word, so close to *animated*, conjuring up images of cartoon characters drawn by an animator who gives them whatever life they appear to have.

Animist is also a word freighted with colonial prejudice. The late-nineteenth-century English founder of anthropology, Edward Tylor, said that animists were "primitive"; they were people who believed in "spirits," or nonmaterial entities. To Tylor, the Anglicans of his time were as animist—and as primitive—as Melanesians or Bantu because they too believed in a spirit (of God). Tylor, in other words, thought animists focused on the supernatural. Because he thought that only the material world is real, to him animists were like children, confused about what is and is not alive.

Our ideas about animism have not progressed much beyond those of Tylor. Animism remains a quaint, outmoded view, and animists are misguided folk who stubbornly cling, despite all modern evidence, to their idea that inanimate objects are alive.

Some try a kinder, gentler approach to animism. They point out that the word *animism* comes from *anima*, Greek for "soul" or "psyche" or "mind." Literally, then, animists are people who view all of nature as "ensouled." But this word too falters on a Western idea of matter as divorced from mind or spirit. In its two simple syllables, *ensouled* implies that the soul has been put there from the outside, instilled after the fact. In the modern Western paradigm, matter is the "fact," while soul is something radically different—the absence of matter. Soul, or spirit, originates elsewhere— in the heavens, in some other dimension, or as some pop spirituality theories proclaim, in the Pleiades—but it always comes from above, never from below. To most Westerners, whether nihilists or New Agers, spirit does not and cannot arise from Earth.

But look closely at that word *anima*: in its original sense it meant "the life within." *Animism*, then, could mean a world that is *alive*, not merely "enlivened," a world where life emanates from matter, where all visible things participate in creating one another just because they "are." In this view, which is my own, nature is personal; it is a place where trees or birds might say hello and where the great community of spirit includes lizard, lion, and lichen as well as the rocks and land that shape and nourish us all.

What does it mean to say that nature is personal? In the broadest sense, it means choosing a different paradigm from the Western idea of nature as machine. It means seeing more continuity than discontinuity between humans and other creatures. It means holding the possibility that all things on Earth, even rocks and mountains, have their own will and intention. For don't rocks too move from place to place, however slowly, shaping the world as they go? Where I now live, in Boulder, Colorado, megaliths of tilted sandstone define the western edge of town, forming both the geology and the personality of the place; if they look solid and still, they certainly are not so across geologic time. "Everything dreams," writes science fiction author Ursula K. Le Guin. "Rocks have their dreams, and the earth changes."

And don't plants use humans to carry out their will, as Michael Pollan argued in *The Botany of Desire*, hitching rides from us to spread their genes or cajoling us with their beauty, as in tulips, or their chemistry, as in marijuana? Pollan may not say that Earth is alive, yet anyone who sets out to tell a story from "a plant's-eye view of the world," as his subtitle announces, has made a mental leap away from mechanism and toward perceiving nonhuman others as having their own will and intention and treating them as more like than unlike ourselves.

And here lies the crux of the matter: to say that nature is personal may mean not so much *seeing* the world differently as *acting* differently—or, to state it another way, it may mean interacting with more-than-human others in nature as if those others had a life of their own and then coming to see, through experience, that these others are living, interacting beings. When nature is personal, the world is peopled by rocks, trees, rivers, and mountains, all of whom are actors and agents, protagonists of their own stories rather than just props in a human story. When Earth is truly alive, the world is full of persons, only some of whom are human.

⌐

A. IRVING HALLOWELL was an anthropologist from the University of Pennsylvania who lived among the Northern Ojibwe of Canada during the

1940s and 1950s. Hallowell's writings are leading scholars of religion to challenge Tylor's old ideas about animism because Hallowell found that Ojibwe thinking simply doesn't fit Tylor's categories. While Western scholars viewed Ojibwe as believing in "spirits," the Ojibwe saw themselves as relating to "persons" of many kinds—tree persons, stone persons, cloud persons, dream persons. Trying to take seriously the Ojibwe cosmology, Hallowell coined the term *other-than-human persons.*

One day Hallowell asked an Ojibwe elder, "Are *all* the stones we see about us here alive?" The old man looked around, considered a long moment, then said, "No! But *some* are." The answer is important. Stones could be related to as "persons," not because all stones have a "soul" or "spirit" but rather because some stones *act* like persons: they relate to others. Naturally, some stones are more "personable" than others.

The concept is not hard for even skeptical Westerners to follow: Hiking through a forest, don't we notice our attention drifting to some trees more than to others? Most of us interpret that moment of noticing as the human mind choosing which trees to focus on. But what if the opposite is also taking place? What if trees, or at least some of them, have the ability to draw our attention toward them?

Notice the difference in the quality of the experience. The moment of noticing can be the act of an isolated human mind. Or it can be a moment of relationship, a meeting of two.

⌒

WHEN DID HUMAN societies, and especially my ancestors in the Western world, stop listening to the whispers of trees and water, moths and mud? David Abram is an environmental philosopher who suggests it took place before Christianity, before Socrates and Plato. In his view, becoming literate led our ancestors to forget their cousins the animals and to seek wisdom instead in the words of humans alone.

In nonliterate societies, he says, humans focus their attention on the animals and plants, weather and tides of the natural world. Plenty of reading takes place, but it is a type forgotten by literate people, the reading of signs and tracks. To people who hunt for their food, each paw print in damp soil tells a story—how the rabbit paused at this point, avoided the coyote at that point—and provides clues about where a rabbit can be found for dinner. Each leaf whispers news of wind and weather and of the creatures who brushed by it minutes or hours before. Reading the wisps and wiggles of cloud on the evening horizon helps to plan the next day's work.

The sense that the whole world is alive, says Abram, is lost once people transfer their attention to books. "It is only when a culture shifts its participation to these printed letters that the stones fall silent. Only as our senses transfer their animating magic to the written word do the trees become mute, the other animals dumb."

Abram could bemoan the advent of literacy and gaze with nostalgic eye on exotic, nonliterate societies, but he does something much more useful: he uses modern people's experience of literacy to demystify indigenous cultures. He makes a radical suggestion: that modern Westerners are animists too, at least in one important way.

This type of animism is taking place right now, in this moment. A reader gazes at words on a page, and little marks of black are transformed instantly into words spoken by a voice in the mind or images arising on an inner screen. Reading a novel, we might picture the characters, feel their feelings, identify with or loathe them, and at the story's end we may be reluctant to say good-bye. We feel as if we have experienced their lives.

So much emotional intensity! And yet we are engaged only with small black marks printed on a leaf of paper. To a nonliterate person it could look like magic—gazing intently at a leaf and taking away stories! Reading makes leaves come alive. For Westerners, reading produces the experience that Abram says is common to all animists, of our senses being engaged so fully with something "inert" that it speaks to us.

Some might argue that a book does not have inherent life: the marks of black ink on its pages do not speak, and only what the imagination of the reader brings to them causes them to communicate. This is the same logic by which we are able to consider rocks inanimate, as if the only life they have is what the human imagination ascribes to them. Yet in relation to a book we can see how specious this argument is: the words on the page are arranged in patterns of grammar and syntax, patterns that we learn to recognize only through the hard work of learning to read. In mastering sentences and paragraphs, we gain access to a different world, the world revealed in a book. We can see its skies and valleys, streets and buildings, hear the voices of its characters or the beat of its surf, smell the grime of its industry or the heady air of its mountains. And once we have experienced its world, we are unlikely to consider that book any longer as a simple object. Ask any bibliophile who has had to move her library across a town or a continent: life does not feel ordered until her books are unpacked and placed again within reach. The books are not objects but friends, each disclosing a piece of meaning, evoking particular feelings and thoughts. We do not doubt it: books speak.

In a similar way, argues Abram, the natural world speaks to those who take the time to listen or read—who parse the grammar of clouds and moisture, who follow animal tracks or leaf shapes in pursuit of a meal or a medicine. One need not postulate a Creator, as an author, in order to contact the aliveness of the sea. Those who have trekked in the desert, with its magnificent sweeps of earth and sky, do not doubt it: rocks speak.

For just as unseen worlds unfold to those who read a book, so worlds hidden to hurried sight unfold to those who choose to spend more than a few moments cultivating their relationship with nature. Paying attention is key: we interact with the other when we allow it to engage our attention, when we "read" it with absorption, as we would a book. The ficus tree in the office cubicle or the oak planted in the urban sidewalk offers un-dreamed-of wonders to those who pay attention. Just because to literate people reading a book is unremarkable, available to anyone who can learn the alphabet, it is no less magical. Among my people, children are taught to read books; among some other peoples, children are taught to read trees.

～

IF WORLDVIEWS ARE human approximations of nature rather than nature itself, then to mistake one's worldview for reality is to repeat an error com-mon to religion: the error of idolatry. We tend to think of idolaters as those who build statues, like casting a golden calf and bowing down before it, as related in the book of Exodus. But while the ancient Hebrews worshipped something physical, an act that was forbidden, the greater error might be one of perception: mistaking a human creation for reality itself. To confuse limited with unlimited is to try to squeeze ultimate reality into the confines of a human model.

One needn't be religious to repeat a religious error. The seventeenth-century model of nature-as-machine, or mind as separate from matter, won the worship of the Western world for four hundred years. So enchanted have we been with its gleaming hide that we have labeled as "supersti-tious" or "ignorant" anyone who chooses a different worldview.

To return to the Zen axiom for a moment: "If you meet the Buddha in the road, kill him" is a powerful caution not only against trusting too much in human authorities but also against mistaking an icon for reality itself. Our seventeenth-century view of nature, in which we have placed so much trust, may be a Buddha in need of slaying. What might we see if our eyes were no longer shadowed by his huge frame?

～

ANYONE CAN CHOOSE a worldview based on relationship; it is open to people of any faith or none at all. One need not believe in spirits or convert to an indigenous religion to count trees as friends. No trance states are needed, no drumming, no psychoactive drugs. No rejecting of one's present faith is required, unless that faith is opposed to seeing the divine in the natural world—and I know of none that is (though fundamentalist forms of theism see larger barriers between God and the world than do the moderate forms of these faiths).

Likewise, no rejection of science is required, although one's worldview might shift, as certain branches of science have already shifted, away from the view of nature-as-machine that still dominates most of modern life. When the world is made of relationships, science's basic commitment to testing through experience can remain central and may even be strengthened as old Buddhas are slain by new discoveries. In my experience, meeting the bald eagle—as well as scores of other meetings with plants, animals, and rocks—did lead me to a new kind of faith, not in the sense of accepting a new dogma but rather in the sense of discovering a guidebook that made better sense of the territory.

And in this alternate guidebook, trees are worthy of attention and birds may become companions on the journey. The pilgrimage is toward nature and the practices simply ones of cultivating relationships with those who are nearby—the oak outside the window, the soil that grows new life, the grass in the crack of the urban sidewalk. These are everyday practices that can be engaged in by people of all religions, people of no religion, and all who count themselves "spiritual but not religious": practices of respect, careful attention, empathy, and kindness.

The simplest practice is taking time just to be in nature. And I don't mean packing your bags and traveling hundreds of miles away to visit some pristine wilderness. Connecting with nature need not mean leaving home, at least not in the geographic sense. I met the bald eagle on a journey, that's true, but not in some "untouched" wilderness. Perhaps the idea of traveling elsewhere to connect with nature springs from Euro-Americans' own heartbrokenness—usually repressed—at having cut down the forests and conquered and banished the people who lived here before, people whose relationships with nature were livelier than our own. For the most part, modern Americans can't find the wild within or around us, so we go elsewhere to find it.

In a worldview where "people" and "nature" are separate, nature can reside only apart from humans; cities are "spoiled," corrupted by technology. If it's in our backyard, we don't call it nature, we call it a garden. In my experience, places unpopulated by humans do sing with their freedom from human tread, and they can indeed evoke a different feeling in us than does nature that is weeded and mowed and trimmed—thus the need always for wilderness—but the nature in our own backyard has equally powerful insights to offer and will also lead to the mystery that we seek.

Perhaps most striking of its gifts is friendship. Just as we receive more sustained pleasure from keeping company with our dog than from a momentary glimpse of a wolf in the wild, so also the plants, birds, and rocks of the ground we inhabit offer a more sustained experience of friendship; they live in close proximity to humans, and so they are available on a daily basis to watch and wonder at, to learn to know as intimately as we would family members, and to develop a sense of kinship with over time.

And if we open ourselves to respecting as friends the trees or plants or rocks in our backyard, we may find that we become travelers anyway, but of a different sort: cultural pilgrims at home. We may become willing to venture outside the bounds of everyday belief in the solidity and separateness of things. We may practice suspending, if only for a few moments at a time, the idea that nature works like a machine, and we may hold instead, if only for a few moments at a time, the idea that other parts of nature have their own awareness, their own will and point of view. We might become willing to experiment, to try things that we have been told are impossible, to risk looking silly in order to follow a feeling or a hunch. And if we do all that, eventually we have stories to tell—as any returned traveler does—about the ways in which the world opened to our explorations, responded to our questions, and offered hospitality and gifts beyond any we could imagine.

~

MY OWN JOURNEY toward listening to nature began many years before meeting the eagle and continues to this day. Although for much of my life I experienced animals and plants as having a lot more to say than humans gave them credit for, not until recently did I see the much larger implications of what I had thought of as private spiritual encounters. These meetings nudged me toward a worldview that offers tools that we, with our habit of treating nature as dead matter, sorely need. It is a new worldview with old, old roots, and because it emphasizes relationship and respect, it might actually help us solve the enormous ecological puzzles we now face.

But when I think of the journey I have taken, it is the bald eagle I met more than fifteen years ago who most clearly rewrote my guidebook. For when the eagle made a beeline for our car then flew in tight circles directly overhead, staring down with sharp eyes, I was forced to see something I had never seen before. I saw a friend—one who had heard my call and who had chosen, out of free will, to respond. And, like human friends who respond, the eagle changed me, not only lifting my spirits but in the long run shifting my worldview as well.

My conversion to this new faith was not completed overnight, but after that meeting on Lopez Island, it was well under way. Step by step I was leaving behind a world in which humans alone possess the ability to communicate. I was moving toward a world in which all things, by virtue of existing, are permeated by life, are actors and cocreators in shaping the world, are bearers and bringers alike of intelligence, of the Great Mystery. After all, the other creatures preceded us by millions, even billions, of years, so in simple evolutionary terms it is more likely that we borrow of their intelligence than the other way around. "Birds were fully evolved by the time we stood upright," says Judith Irving, who made the film *The Wild Parrots of Telegraph Hill,* so it's "better to describe human behavior as avianomorphism."

After fifteen years, time has only heightened my appreciation for what the bald eagle accomplished. Then, I was amazed that the eagle heard my thoughts. Now, I accept that ability to hear and be heard as normal, as the very fabric of life, stitching together beings throughout the world. The miracle now seems to be that the eagle came in response, flying out of his or her way to greet me.

For when I look back, I see that my request was very nearly a command. The nerve I had, calling eagles away from their daily lives just so I could get a good look! No person, human or other, wants to be ordered around, made to conform to the wishes of another. A Buddhist teaching says, "To every being its own life is precious," which is usually taken to mean that all creatures get caught up in preserving their own lives. I like to think it means something else as well: that every being wishes to be free, not enslaved to another's purpose. I have never since called another creature to show up just for my convenience.

Yet my audacity was honored, and perhaps this too is a quality of friends. For if someone dear to us makes an urgent request, won't we do what we can to respond? At that moment in my life I was in sore need of friends, and this eagle heard my request and chose to respond.

When I got home from Lopez Island, though, I wasn't yet thinking in these terms. I knew only that I was grateful to the eagle and that feeling grateful was a welcome change from the despair that had shrouded me. I was content, for the time being, just to notice and wonder at the experience. What the eagle had brought to pass was a miracle, however I might slice it.

All that would flow from that encounter I couldn't quite imagine: greater appreciation for the mystery of life pervading all. More delight and less loneliness. A deeper listening to the body, to the physical world, and thus a greater hearing of spirit. The inner peace that can arise when we lay down arms against the self, and the ecological peace that grows from laying down arms against nature. More harmonious ways of understanding the self and community. Most of all, a spirituality of this world, in which prayer becomes simple greeting and ethics is encompassed in relationship.

All I knew that day was that in responding to my call, the eagle opened up new avenues for knowing; it revealed connections available every day to those who care to notice the birds, trees, rocks, or plants beside their house or along their street. For me a new way of being would grow from that encounter, a way of being in the world that had the potential to heal our perceived split between humans and the rest of the creatures on Earth and so head us ecologically in better directions than our old worldview had prepared us to do. It was a solution as simple—and as challenging—as relationship: the solution of widening our circle of friends.

1

Cut-Leaf Weeping Birch

O dear to me my birth-things—All moving things, and the trees
where I was born—the grains, plants, rivers . . .

— WALT WHITMAN, LEAVES OF GRASS

F ROM THE SMALL cardboard box in my closet I pull a single photo—a
four-inch square of faded color showing a tall, weeping birch tree,
its leafless winter branches a graceful spiderwork of wispy white. The air
is hazy, as the previous night's ice storm has condensed now into an icy
fog. Pale background light suggests a watery sun trying against all odds to
spread rosy hues over new-fallen snow.

I remember taking this photo—around eight o'clock one morning just
before heading off to high school in Archbold, a tiny town in northwest
Ohio. My favorite tree, planted just beside the house, looked breathtaking
sheathed in ice.

Next to this photo in the same cardboard box is another one of the
birch tree, this one taken in autumn. Bright gold leaves and white trunk
shimmer against a cerulean frame of October sky.

Fifteen years later, when I was thirty-three and just beginning the journey of listening to nature, the birch tree became a friend at a deeper level.
I was living two thousand miles away from it on a hillside in Oakland,
California, when I was startled one evening by something I'd never before
experienced. I'd recently had to get used to a lot of changes. At Christmas
I'd fallen ill with a flu and never fully recovered, and for half a year now I'd
scaled back my life—less editing work, slower doctoral studies, going out
with my husband only rarely. Now it was June of 1990, and I'd just been
diagnosed with chronic fatigue immune dysfunction syndrome (CFIDS).

I dragged myself through each day feeling as if twenty-pound weights
were strapped to each limb. Each step took considerable effort: I could feel,
but not see, that our hillside flat flowed ever so slightly downward because
walking across the living room floor took less effort in one direction than
the other.

One evening as I headed downhill on that floor just after dinner, an
image popped into my mind. It was the birch tree, and in the space of a moment the tree was as present inside my head as the sofa was present before
my eyes. I sank onto the sofa and closed my eyes. What was happening? I
hadn't thought of the birch tree for years and had no reason to be thinking
of it now.

There was no announcement, there were no words. The tree simply
rose in my awareness, its tall, graceful image strong in my mind—stronger
somehow than a memory. The presence remained vivid for a few moments
then gradually faded. A feeling of quiet gravity, almost sadness, had accompanied it.

Two weeks later I received a call from my older brother, who still
lived in Archbold. We talked of the failing health of our aging parents,
the estate sale that was soon to come. Then Bro added slowly, "Well, I'm
afraid we're going to have to cut down the birch tree. It's got a disease or
something."

Ah, so that was it. The tree had come to say good-bye. I hung up the phone and stopped for a moment to focus my attention on the birch tree again. How was this possible? I didn't know, but I felt grateful.

That autumn, at the sale of our parents' house, I picked up pieces of freshly cut birch limbs to carry home to California. One I burned on New Year's Eve some years later, gazing quietly at the flames that curled and flashed around the white skeleton of my old friend. In the new year I gathered the ashes carefully into a grocery bag and took them to a ceramics class. There my teacher helped me stir them with other minerals into a glaze, and I painted the mixture onto a slim vase I'd recently sculpted. Today, as I write, the vase sits on a cabinet across from me, its sweeping curves finished in speckled cream.

↶

I RECALL THIS incident with the birch tree as the first time a nonhuman being visited me. But it's probably more accurate to say that it was the first time I was able to receive such a visitor.

In my childhood there was a noticeable lack of mystery, for the sober German American Mennonites to whom I was born valued hard physical work, plain clothes, and above all, plain thoughts. Their sometimes severe practicality fit well in our small town, where cinderblock factories swallowed people into their deafening noise five days and nights a week, lawns were mowed weekly, and shrubs were barbered to smooth green skulls. Surrounding the town, lands that for millennia had harbored a great, deciduous wetland now nourished tidy and well-managed squares of corn and soybeans. European settlers in the nineteenth century had razed more than a thousand square miles of forest and planted drain tiles to take away the water so they could plow and plant their crops, and the flat earth looked now like what it was: a denuded landscape that in summertime was lush and green but in winter appeared exposed, a monochrome gray-brown. The only remnant of the Great Black Swamp of northwest Ohio—that loamy stew of water, trees, meadows, and microorganisms—were tiny stands of woods tilted here and there to the horizon like tombstones to forgotten ancestors.

Throughout the plain-thinking years of my childhood, I sat through endless days of school, bored with lessons yet unable to disappear into fantasy. I couldn't think up stories like other kids did. I didn't know how to make up games. I was sure I had no imagination. Yet pieces of mystery continually stretched toward me. Through the center of town, giant trees arched over Main Street, the tips of their branches touching overhead, and I biked under their bowing limbs, reveling in the green canopy. The town library, just across the street from the elementary school, harbored myriad treasures of words, and, lost in the pages of a book, I discovered worlds undreamed of. Books and trees: these were the mediators of mystery in my childhood, and though as I grew I devoted more time to books than to trees, still my connection to trees remained strong—especially to one tree, the birch in our yard planted five years before my birth.

As a child I loved to stand amid the dangling threads of its branches, next to the slim white-and-black trunk. Its young bark was thin and crinkly as tissue paper and revealed to my toddler fingers the joys of peeling. In junior high, sent to rake the birch tree's generous fall of leaves—tiny crackles of serrated yellow—I would rake and daydream, sometimes arranging the leaves in the rectangles and squares of a floor plan, with windows and doorways laid out as in blueprints. And when I was in high school, the tree, now thirty feet tall, witnessed my longing as I sat beneath it on summer afternoons, immersed in books, reading everything from romance novels to C. S. Lewis. I can't say I shared secrets with the tree, but I did feel different—calmer, a little more confident or clear—after being veiled for a few hours under its leafy fall.

↶

IF BOOKS AND trees mediated mystery in my childhood, it's probably no accident that when, as an adult, I became ready again to receive news of a world beyond ordinary logic, books and trees delivered it gently to my awareness. I am tempted to think that it began with a book I was copyediting around the time of the birch tree's visit. It was one of the most beautiful books I'd ever worked on—and one of the shortest, since each left-hand page carried only a sentence or two, with a huge, full-color photo splashed on the right. I found the words soothing, and their sparse scattering was all that I, with my illness-addled brain, could edit at the moment. The photos too felt nourishing, each one a close-up marvel of flower petals and stamens, leaves and bark, each dazzling photograph calling to mind those larger-than-life blooms by Georgia O'Keeffe.

The sentences in the book were written by a woman who said she had cooperated with plant spirits to grow incredible bounty in the Findhorn garden in northern Scotland. I was intrigued; could it be that awareness resided in nonsentient forms of life? On one page was a huge fuchsia beaded with dew, and facing it a thought that the writer and gardener, Dorothy Maclean, had received from fruit trees: "Nothing is worth doing unless it is done with joy; in any action, motives other than love and joy spoil the results. Could you imagine a flower growing as a duty and then sweetening the hearts of its beholders?"

A door inside me opened the moment I read those words. I recognized myself—doing out of duty for most of my life, with far too little sweetness. At the time I could hardly imagine what it was like to do things simply out of love and joy. That day I opened myself to a little more sweetness of heart. Could it be that flowers had their own point of view? Even now, when I gaze at a flower, I recall these words and am brought for a moment into the presence of joy, not just the joy of beholding the flower, the joy moving from me toward it, but the joy present in the flower itself, offered to the world—to me—as a gift.

⌐

MY NASCENT OPENING to a world beyond ordinary logic was speeding up because of illness, and especially because Western medicine could not relieve my terrible symptoms. Despite weeks of prescriptions ranging from antianxiety meds to antivirals, I could find no easing of muscle aches or calming of a frantically racing mind, no relief from the hours of insomnia every night or the mental fog isolating me by day as if behind a Plexiglas wall, no softening of lymph nodes hard as walnuts, and above all, no reprieve from the crushing fatigue, the feeling of lead weights pressing down on every square inch of my body.

The person I had known as me was dying, her once-defining abilities now fading and disappearing. First to go was music. Before getting sick I'd been a serious classical musician, teaching a few private lessons in both voice and oboe. Now, blowing air through a stiff oboe reed would have been tantamount to bench-pressing three hundred and fifty. My short-term memory eroded to such a degree that at times I lost the ability to read, no longer able to connect words at the top of a page with those at the bottom.

The continuing pain and distress pushed me, in 1990, the first summer of the illness, onto the treatment tables of two alternative practices I wouldn't have dared to try otherwise. Desperate for relief, I walked sluggishly down to the end of my street, where a Chinese doctor, a woman in her forties, practiced acupuncture. As she threaded needles for the first time into the top of my skull, she watched me closely, asking anxiously, "You okay? You all right?" Yes, I was all right, but the way she jabbed the needles in my legs induced electric shocks so strong that my body nearly jumped off the table. (I learned later that she gave one of the strongest treatments in town, twirling needles to an extent considered unmerciful by others.) When she took the needles out, I was unimpressed with the results of the treatment—until I walked slowly back up the street and noticed, for the first time in months, that leaves on trees were actually *green* and *alive*.

The Plexiglas wall had vanished! For a few precious hours I remembered what it was like to think more clearly and feel like myself again. Within hours the wall rematerialized, but there had been no doubt: behind the veil of fog, a person with abilities I used to have was still there, waiting. I was intrigued; could it be that acupuncture, in treating the body, also treated the mind?

In addition to acupuncture, I was regularly visiting the tiny meditation room of a church across the bay where a woman of sixty with clear, bright eyes and a wry, bemused smile offered spiritual healing. As I lay on a massage table listening to a recording of Tibetan bowls being struck like gongs, their wash of echoes bathing the room, she would circle the table reverently, placing her hands on my head or legs and sometimes sweeping the air over my body with her palms. During those treatments I would contact a well of calm that most other days was unavailable. The healer seemed able to see trouble or distress in the organs of my body, and from time to time she would place her hands on my abdomen and grow more deeply quiet, as if listening. "They all like to do their jobs well," she said. Who knew that organs had a point of view?

BY THE TIME I resorted to alternative medicines, I had been a feminist for at least a decade and a spiritual seeker for what seemed like forever. But now I was forced to practice both feminism and spirituality with a level of commitment undreamed of in earlier years. Fortunately, this was not hard, at least not philosophically; illness was showing them to be not two separate paths but one. The courage I'd gained from feminism—risking trust in my deepest inclinations—seemed no longer just *a* path toward healing; it was apparently the only path. The severe necessity of illness was forcing me to pay attention to the effect of every activity or emotion or interaction on well-being—to listen to my body not just in a vague or general way but with unwavering attention.

I began testing all decisions—even mundane ones, like which herb tea to drink for breakfast—on a scale of breath and muscle and heartbeat. An internal yes or no would arise in a feeling of lightness or heaviness; things tending toward heaviness, I learned, sent me deeper into illness. If people or activities depleted my energy without replenishing it, I had to limit my time with them, no matter how long I'd known them or how deeply I cared for them. I was subordinating all of life, not to a scripture or to a spiritual teacher, but to my own health. It was a fierce feminist practice of honoring a woman's body and a woman's needs—the woman in this case being me.

Words of poet Audre Lorde had long graced my desk: "Out of my flesh that hungers / and my mouth that knows / comes the shape I am seeking / for reason." These lines came from a love poem, yet they seemed to illuminate illness as well. I was learning another side of the erotic—flesh hungering for health—and this hunger too, it seemed, could define a new way of being in the world.

It felt scandalous, this commitment to putting the body's needs first. By that time in grad school, I'd studied the history of Western Christianity, which was marked by a deep suspicion, sometimes hatred, of the body. The ascetics of the fourth and fifth centuries had fled a disintegrating Roman Empire by moving out of their towns and villages and into the desert, renouncing sex and marriage to take up spiritual contemplation; by medieval times the truly devout controlled their bodies by controlling their food intake. In these and many other ways Christian history spoke again and again of the body and its urges as a problem to be remedied—by fasting, by abstinence, by renouncing physical pleasures of all sorts.

Not just the body but the whole self was to be denied in many versions of piety surviving down to our time. As a child in Sunday school, I'd been taught a little song to the tune of "Jingle Bells" about the proper order of things: "J-O-Y, J-O-Y, this must surely be: Jesus first and Yourself last and Others in betwee-een; J-O-Y, J-O-Y, this must surely be: Jesus first and Yourself last and Others in between!" The song echoed millennia of Christian teachings about finding heaven by denying the self.

Now chronically ill, I had to practice the opposite, renouncing not the self or the body but the kind of religiosity that labeled self-care as selfishness. Audre Lorde, battling cancer, had said, "Caring for myself is not self-indulgence, it is self-preservation," adding bluntly, as a black lesbian feminist, "and that is an act of political warfare." Healing from illness, I was finding, required spiritual insurrection as well.

~

NOT THAT CARING for the self is any easier when ill; it may in fact be harder just because living with illness is, by definition, harder. But, reduced to discomfort or pain, one is likely to seize more quickly on anything that can possibly lend relief. And so, robbed of the ability to do things that had previously defined me, I resorted to doing the few activities that were still available and seemed to help. Chief among them was resting.

I'd already spent a decade practicing a relaxation exercise learned just after college in a yoga class. At the end of each class the teacher led us in a ritual, relaxing each part of our bodies as we lay on our backs, inert on our mats. I didn't know that this relaxation exercise was the yoga pose called *shavasana*, or corpse pose; all I knew was that when weakened limbs ached and mind raced uncontrollably toward destinations legs could not travel to, doing this relaxation exercise brought mind and body a bit closer together. A state of relaxation was harder to find than when I'd been well—god, was it harder!—for lying still was nearly impossible with muscles and mind made twitchy by illness, but if I could manage even a few moments of utter stillness, the achiness subsided ever so slightly and agitation eased. This was enough to keep me at it.

Not until many years later, when I rediscovered yoga and took it up in earnest, did I realize that I'd been practicing what is considered its most important pose. One can enjoy shavasana as simple relaxation, or one can go deeper, releasing cares, letting go of effort, entrusting the body to gravity, to the earth—preparing, in effect, for dying. It seems no accident that during those years of illness, when the person I had been was dying, the one pose available day and night, the only pose always available, was the pose of the corpse.

About the time I took up yoga again, I went out for dinner with some religion scholar friends who are also meditators. Over dim sum we identified our various spiritual practices, and I mentioned relaxing. They looked at me quizzically, even after I explained that it was a way of settling into my body. Then I remembered—of course! Many spiritual practices try to do the opposite; they help people escape from the body, as if the body were a prison and only the mind or spirit were free. It's a logical goal if spirit is thought to be separate from matter, and if so, spirituality will tend toward the immaterial—toward prayers and pujas designed to rid people of bodily limits and take people out of this world—rather than toward deeper intimacy with the world that is our home. Comprehension finally dawned in my friends' eyes when I explained, "All my life I've been trying to learn to trust *what is*."

&

BY THE TIME I became ill, at Christmas 1989, both Christian feminist theologians and Goddess the*a*logians had already spent decades resisting organized religion's flight from the body. Instead of looking, as did patriarchal religion, to a transcendent God-in-the-sky, a God of spirit separate from matter, they focused on the God or Goddess present in matter, on themes of immanence and the indwelling of the divine.

Yet whether they recommended a God or a Goddess, I had found myself restless with all of them—even before I fell ill with CFIDS and began the long work of recovery. I pined for a picture of reality—of the divine—that was not limited by the human form. It was just too taxing to conjure up the correct image, whether Christ or Goddess, as a focal point for faith, always relating the here and now to some sacred story that had happened long ago or far away. My spiritual restlessness had landed me, a half dozen years before falling ill, at Pacific School of Religion, a progressive Protestant seminary in Berkeley.

During my first semester in Berkeley, in 1983, curious about other religions, I'd signed up for a course in the Upanishads. The professor was a doctoral student in Indian philosophy, and she taught us in the traditional way of ashrams, introducing us to the ancient texts through the lens of Shankara, the ninth-century founder of the school of thought called Advaita Vedanta.

Tat tvam asi, Shankara taught: "Thou Art That." It's the only Sanskrit I remember from the class, but it summed up the Upanishads, or at least Shankara's interpretation of them. The infinite, changeless, forever-bliss of the Absolute—what we always look outside ourselves to find—is not outside us. Shankara taught a radical idea: the Absolute can be found in your own deep center, your Self. "It's closer to you than your thumbnail," the professor said, paraphrasing the Upanishads. What is inside us is not different from the eternal.

> As the same air assumes different shapes
> When it enters objects differing in shape,
> So does the one Self take the shape
> Of every creature in whom [it] is present.

But if the Self is not different from the eternal, neither is it identical, exactly. *Advaita* means "nondual," and *nondual*, the professor said, means "not different, not the same." I liked that thought—the Absolute is within us yet bigger than us too. The world is not its measure, yet we and the rest of the world are its evidence.

At the end of the semester I discovered I was a nondualist. God, for me, was no longer outside the world, as in monotheism; God was wrapped up in the world. Look at a tree? We're looking at God. Look deeply inside the self? We're looking at God there too. Focused less on human figures, nondualism seemed paradoxically more welcoming and spiritually inclusive. And it fostered a sense of inner peace, of harmony, as if laying down arms against oneself, for as the Katha Upanishad also says, "Eternal peace is theirs who see the Self / in their own hearts."

A couple of years after taking the Upanishads course, and a few years before falling ill, I revisited the Mennonite seminary in Indiana where I had studied before transferring to Berkeley and sought out a professor with whom I'd worked closely. "What have you learned since you've been gone?" he asked. My mouth opened to speak words I hadn't planned. "I've transferred my center of authority from the outside to the inside," I heard myself say. He smiled quizzically.

In the years before I fell ill, nondualism and feminism together had been at work in me, healing a trust in the deep self ruptured by years of steeping in a Christian brew of original sin and human fallenness. Together they had provided a philosophical basis for relying on the self instead of on outer authorities, and together they had built a foundation for inner work, for if God is found in your own soul, then soul-work takes you there, and bringing forth what is deepest and most precious inside you is divine work. And now, when I was ill, they worked together again to provide a foundation for listening to my body and following it as my teacher.

Not, of course, that I understood this at the time.

⌒

AFTER THE BIRCH tree showed up to say good-bye and the weeks of chronic illness dragged into months, forced to listen to my own body, I found myself listening to other parts of the physical world as well. The flat I lived in with my husband opened out to a deck and garden shaded by an immense redwood tree on the other side of the fence, and on days when I was unable to do much more than lie on my back and gaze upward, I would choose a spot on a bench and simply stare up at that tree, sometimes for hours on end. The tree was a coast redwood, its silhouette dark against the bright sky, its thick, sprawling limbs drooping generously toward the ground. I knew from picking up some of its wind-downed branches that the needles were thick and flat, a few branchlets ending in small round cones. I noted its colors in different kinds of weather—how an overcast day with indirect light might brighten the rusty color of its shaggy bark—and I watched its conical tip wave high above neighboring houses in the breeze. The tree eventually felt like a friend, a familiar; simply noticing all these details over time made it so.

Another plant that felt like a friend was a scarlet bougainvillea draping voluptuously around our kitchen window and coloring the white cabinets rosy with its reflected light. When, in late 1990, in the second winter of my illness, a severe frost hit the Bay Area and the bougainvillea's leaves rotted, I felt joy depart from the kitchen. In January the withered tangle of woody vines had to be cut down. I walked outside to visit the stump at the base of the house, feeling sad. Then a thought occurred to me: *The healer places her hands on your body, and you feel better. You could just place your hands around the stump.* It was ludicrous. I refused. But I couldn't make the thought go away. Glancing over my shoulder to make sure no neighbors were watching, I bent to the ground, acting as if I was just inspecting the plants there, and placed my palms furtively around the stump. I tried to quiet my mind, to not feel stupid for doing something so simple.

And into the middle of my arguing mind rose a sudden, sharp image—the stump with two green shoots poking out from it, a larger one on the upper left, a smaller one on the lower right. Amazed at the clarity of the image, and feeling now only slightly less stupid, I stood up, glanced around again, and returned to the house.

The rest of winter passed. Spring came, bearing pink buds on plum trees and purple hyacinths beside sidewalks. Yet the bougainvillea remained dead. Now I knew beyond a doubt that I had imagined the whole thing.

Weeks passed before I ventured out to that side of the house again. This time it was to retrieve the bird feeder, which had been knocked to the ground by hungry, nesting house finches. I stepped through a tangle of ivy to pick up the feeder, where it lay next to the dead bougainvillea stump.

My eyes widened. From the dead stump two green shoots were springing—a sturdy one several inches long on the upper left, a shorter one on the lower right.

        ⌐

SO WHY WOULD I feel so self-conscious placing my hands around a dead stump? I lived in California, for pete's sake, where people did crazy things every day, often for the sake of spirituality. Or so it was said, though I saw no evidence of it in my neighborhood, where people headed out at five or six or seven each morning for schools or offices or Silicon Valley labs; no one as far as I could see spent their busy days nursing dead stumps. I was afraid of being thought insane—not just for using my palms in healing but for doing it to a *plant*, because as everyone knows, the only things plants need to survive are water and a good piece of dirt. Still, that day under the kitchen window I tried offering something less tangible—some kindness, some healing and life-sustaining warmth, something humans call "spiritual." I was treating the plant as if it had soul or spirit—as if the bougainvillea were a friend.

        ⌐

THE TWENTIETH-CENTURY ANTHROPOLOGIST Claude Lévi-Strauss told a parable about the original inhabitants of the West Indies. Some years after Columbus made land in 1492, he said, while Europeans were sending out commissions of the Inquisition to put the Indians on trial to find out if they had a soul, the Indians too were testing Europeans; they were practicing trial by drowning, to see if Europeans' bodies decayed like theirs. (I've never seen any evidence of this; I suspect trial by drowning was a peculiarly European perversion.) Lévi-Strauss's point, during an age when the word *savage* was still used to describe non-Western peoples, was that humans are humans across the board: we all divvy up experience according to our worldviews, and we're all ethnocentric, trying to figure out if others are like us.

But more than fifty years have passed since Lévi-Strauss's parable, and the years reveal another level of meaning. As Brazilian anthropologist Eduardo Viveiros de Castro says, both the Europeans and the Indians may have been ethnocentric, but they were ethnocentric in opposite ways. The Europeans assumed that all the creatures they met up with had bodies, but the "marker of difference," the great distinguishing feature of humans as opposed to other creatures, was the soul; you're human if you have it and animal if you don't. The Indians, by contrast, assumed all creatures they met had souls, and the marker of difference was the body; you're earthly if you have it and ghostly if you don't. "European ethnocentrism," in other words, "consisted in doubting whether other bodies have the same souls as they themselves; Amerindian ethnocentrism in doubting whether other souls had the same bodies." He finds the contrast useful because the Amazonian people among whom he lives, he says, hold the same world-view. For them soul or mind is the given, the universal, the quality shared by humans and all other beings, while bodies are variable; animals, plants, rocks, and humans all have different ones. That is, in a Western worldview, humans and all other creatures share "body" (physicality) but not souls;

in an Amazonian worldview, humans and all other creatures share "soul" (life force) but not bodies.

～

WE TEND TO think of the Amazonian view as primitive, something Westerners left behind long ago and that Amazonians too must abandon as they join the modern world. Our own cultural trajectory veered just after the Middle Ages away from the view that all beings have souls. During the scientific revolution of the seventeenth century the view took hold that nature is a machine and that only humans have souls, and because it has been centuries since we perceived the breathing of the world, people who consider the world ensouled appear to us medieval.

Yet so much turns on this definition of *soul*. To Westerners today— those who believe in it—the soul is a disembodied spark of life housed in and often in conflict with the body. The well-known British theologian N. T. Wright puts it perfectly: "a disembodied entity hidden within the outer shell of the disposable body." It might surprise modern believers in the soul to learn that Plato, for all his ambivalence about the relationship of soul and body—to him they were always distinct—would never have dreamed that the soul was disembodied. Soul was an animating principle, the source of motion, but it had its own substance—different from that of the body, to be sure, but still a type of solidity; for Plato and his contemporaries, the soul took up space.

I can't help but think of the *Star Trek* race of Trill, who looked like humans but carried in their abdomens a huge, slug-like body called a "symbiont," which was actually a personality with a set of memories different from those of the "host." When implanted in a host, the symbiont and host became a new, merged personality for the duration of the host's life, at the end of which the symbiont was removed, taking with it the memories of that host into its next one. Probably only sci-fi geeks would explain Plato with *Star Trek*—though in grad school plenty of us religion students loved *Next Generation* and *Deep Space Nine*. Yet the symbiont, in being both a material body and an immaterial set of memories, comes the closest I can think of to the substantive soul of Plato. There is another similarity too: in Plato's worldview the soul transmigrated, reincarnating in a new "host."

Plato needed a concept of soul, for in his world of fixed essences, how else could one explain motion or change? And since for him all motion or change signaled the presence of soul, it meant that plants and animals possess soul just as humans do. But this did not imply for Plato a great democracy of being. He said clearly that souls were not all of the same quality; those that were reasonable always reincarnated in men, those that were weak or cowardly in women, and those that were stupid in plants and animals. Still, the idea that animals and plants and all moving things are permeated with soul, as are humans, had an extremely long life in Western history: about two thousand years after Plato—up to at least the sixteenth century—and who knows how long before him. Plato's hierarchy lived on too in what came to be called the "great chain of being," a worldview in which life forms were ranked higher to lower depending on their amount of life or being, with God of course at the top followed by the angels, then humans (men naturally higher than women), then animals, insects, and plants. Rocks and minerals, lacking motion, lacked life; in the great chain of being earth made up the bottom-most link.

The European thinker who did most to crack this chain of being was the seventeenth-century French philosopher René Descartes, who famously said, "I think, therefore I am." It was not his emphasis on mind, though, that eroded Europeans' views of nature. Descartes was a mathematician and experimenter who knew that motion could be understood as a measurable, physical quality—the distance from point A to point B. But in his day attributing motion to *bodies* rather than *souls* was a radical move. Descartes wanted to show that motion was independent of soul or mind (the two were synonyms), but the reigning worldview saw body and mind as so completely fused that Descartes had to take pains to tease them apart. To state the matter clearly, he overstated it: "There is nothing included in the concept of body that belongs to the mind, and nothing in that of mind that belongs to the body." The old Platonic view that the soul can take up space was, of course, nonsense; so too was the idea that any creatures but humans had souls or minds or could feel pain.

Thanks to Descartes and his radical separation of mind from body, modern Western people find it almost impossible to think about nature in anything but mechanistic terms. For if matter lacks mind, then it becomes an object, operating as a machine. Descartes's views resulted eventually in our present conviction that humans, endowed with mind and choice, are radically separate from the rest of nature. As "geologian" Thomas Berry put it, "Descartes taught us that there was no living principle in the singing of the wood thrush or the loping gait of the wolf or the mother bear cuddling her young." Descartes, in effect, killed off the rest of the world, reducing it to "resources" to be consumed by humans.

If there is a villain in this story, it would have to be Descartes. Even so, I harbor a perverse and secret fondness for the guy. I mean, such audacity! By all accounts his audacity came with large doses of self-aggrandizing, but still, don't you have to admire, at least a little, his ability to challenge the worldview of his time so directly?

Yet what begins as freedom can easily end as chains, and so it is that four hundred years later we wrestle with a problem that did not exist before Descartes: how to explain the mind's interaction with matter. The problem is acute in the wake of twentieth-century revolutions in physics, with discoveries that quarks and electrons in the subatomic world are doing strange things that look suspiciously like they have mental properties. Philosophers have been struggling to explain how something immaterial like the mind—notice the starting point here, Descartes's idea that mind is completely outside the body—can interact with the physical world. They posit theories like "emergentism," or the idea that at some point in evolution, inert matter began to develop mind, a cumbersome idea because it raises more questions than it answers. The problem, it bears repeating, begins when you start with Descartes's notion that body and mind are separate and completely different, with body being physical and mind or soul or spirit being nonphysical. Thanks to Descartes, then, modern people, both religious and skeptic, view mind or soul as disembodied, something Plato never would have done. Thanks to him, we are more Platonic than Plato.

While the ancient Greeks needed a concept of soul to explain change or motion, that concept lost its relevance in the seventeenth century, when motion came to be described in physical terms. So much, as I said, turns on a definition. Obviously, we've moved beyond Plato's concept of soul; outside of *Star Trek*, no one today would make a serious claim for a soul that takes up space. Yet our present ecological crisis, our sense of separateness from the rest of nature, is making us strain at Descartes's disembodied concept as well. We see in hindsight just how thoroughly it has prevented us from realizing our connection with the world that is our home.

I wonder sometimes, What if we thought of soul or mind as a life force, as do people of the Amazon? Bypassing the West's ontological split between "material" and "immaterial," a view of mind as life force could move us toward a wider democracy of being. All things, simply by existing, would participate in the great life-stream, everything moving and unmoving alike permeated by spirit. Such a view of soul, being less precise and more poetic, might help us welcome all beings into the great room of spirit. Would our world look different if rocks as well as rhododendrons could live?

ᘒ

ILLNESS REDUCED ME to lying on a bench and staring at a redwood or placing my hands on a dying bougainvillea, but sitting with trees is something I do to this day, just because it feels good. Today as I write I sit next to a cottonwood tree clinging to the gravelly bed of Granite Creek in Prescott, Arizona, a town I visit several times a year to teach in the graduate programs of Prescott College. The first thing I notice is not the cottonwood tree itself but the noise emanating from it. It is August, and an uproar of cicadas fills the air. Before my first summertime visit to this town, I had never heard such a din. Locusts in my Ohio childhood sawed their songs in melancholy rhythms during July and August, but here in Arizona the insects swell in a chorus so blaring they drown out human conversation. The din is conducive to walking in solitude along a creek and sitting to gaze at a tree.

I look at this cottonwood's bark—gray-tan, roughened with deep creases that weave gracefully up and down the trunk. Limbs are lighter in color, with the white surface common to both the birch and willow families. The white on the limbs reminds me of my first tree love, the birch, and I feel at home with this one and its companions lining the creek. It also reminds me of the aspens, resplendent in October gold, that I enjoy in the Rockies near where I now live.

I send a feeling of appreciation to the tree—for its being in the world, for the clean air it provides by soaking up exhaust fumes from the nearby street. I love its heart-shaped leaves, and the way sunlight seeps through them, rendering them translucent to the dance of shadows from leaves farther up. From the trunk and main limbs pour small sucker branches, their leafy clusters identical in outline to clusters of ripened grapes.

If you sit with a tree, like this, simply noticing and appreciating every detail you can, you may find yourself settling deeper into quietness, deeper into your body. And as you move deeper into yourself, you may find that you move somehow deeper into the tree as well. The quieter you allow your mind to become—soaking up details of this particular tree, this sunlight in these leaves at this moment—the more you may find that your awareness moves not toward the minute but toward the whole. By noticing the smallest feature of bark or leaf, your mind may become quiet enough to take in, paradoxically, the entire tree. And then, even more quietly, you may slip unnoticed into the tree itself. You may find yourself exploring what it feels like to plant your roots in a creekbed and remain, just that way, for decades, sheltering a din of cicadas in your branches in late summer and then, when cool nights lengthen, offering up your harvest of leaves.

And if you can settle even more quietly, your mind aware of the feeling of your roots in damp earth and your branches spreading into light, you might find yourself able one day to take in your surroundings as well: grass beside you, houses just up the rise, cars on the street—perhaps the entire landscape present to you with no effort on your part. Then you feel graced with a sense of peace, of fullness, a simple fact of presence.

And when it is time to collect yourself gently from those roots, to withdraw quietly from those waving branches, you might blink slowly, take a deeper breath, and become aware again of this body sitting on this bench. You are beside the tree now, a companion again, a separate body with its separate purposes.

You stretch and breathe again and stand and begin to make your way back along the path. The world is just as it was before—except now you feel refreshed. A small piece of a larger, all-is-well world has returned with you from your visit to the tree, and you are cleaner, simpler on the inside and so can make your way more easily in the outer world than you could before.

❧

MEDITATION TEACHERS OFTEN recommend sitting quietly and gazing on some object to still and clear the mind; we're less accustomed to hearing that advice from scientists. Yet some of the greatest scientific minds claim that they made their discoveries through absorbed reverie, especially when their meditative focus was tinged with love or appreciation. Einstein said, "The state of feeling which makes one capable of such achievements is akin to that of the religious worshipper or of one who is in love." The first time I read that, I had to look at it twice—not that discovery brings a sense of awe, but rather that awe and love lead to discovery.

Looking at nature with compassion was the method of Barbara McClintock, the 1983 winner of the Nobel Prize in Physiology or Medicine. McClintock was a geneticist working to decipher the maize genome at the same time in the 1950s that her peers Watson and Crick were discovering the double helical structure of DNA. Unlike most geneticists, however, who thought of genes as fixed units, like pearls on a string, McClintock watched, puzzled, as maize genes jumped from their supposedly fixed positions to take up other spots on the strand. McClintock's discovery of "transposable" genetic elements inaugurated what Stephen Jay Gould called a second revolution in genetics—the view of the genome (the part of a cell nucleus carrying the chromosomes) as dynamic, with movable rather than fixed genetic parts. It also prepared the way for the more recent discovery that the genome responds to the environment as well as expressing inherited instructions. The implications of this second revolution are yet to be fully noticed by the general public, but as one *Science* writer observes, when it comes to understanding behavior, it means the old nature-nurture debate is truly passé. Environment ("nurture") and genes ("nature") are locked in a tango, and together they dance their way into our behavior.

McClintock often said that in order to understand any organism, you have to "get a feeling for it." In her small maize field she walked meditatively every morning during the growing season, memorizing the smallest changes in each plant from the day before. "I start with the seedling," she said, "and I don't want to leave it. I don't feel I really know the story if I don't watch the plant all the way along. So I know every plant in the field.

I know them intimately, and I find it a great pleasure to know them." She regarded her stalks of maize, she said, with "real affection," watching each as if from the inside—as if, a colleague remarked, she could write its autobiography. Gould observes that hers was the method of naturalists, who typically spend time watching and listening to—and developing appreciation for—the plants or animals or landscape they study rather than, as most molecular biologists do, trying to isolate chemical chains of cause and effect. McClintock's genius lay in applying the method of naturalists to her work in the lab.

Both a naturalist and a contemplative—don't the two often go together?—McClintock in her deep gazing may seem familiar to those who have practiced meditation or gone on retreat in a monastery or ashram. I think of one of her breakthrough moments in the laboratory, when, after some days of feeling stymied, unable to make sense of the tangled chromosomes under her microscope, McClintock took a walk to sit under a eucalyptus tree. She returned to the lab feeling energized. When she looked again through the microscope at the chromosomes, she reported,

> I found that the more I worked with them, the bigger and bigger [they] got, and when I was really working with them I wasn't outside, I was down there. I was part of the system . . . and everything got big. I even was able to see the internal parts of the chromosomes. . . . It surprised me because I actually felt as if I were right down there and these were my friends.

The process of looking closely at the chromosomes led her into a feeling of unity with them, which led in turn to more accurate understanding of how they operated, seeing them as clearly as if she were moving among them. What is remarkable about her form of contemplation, and what makes it accessible to nonscientists, is that, as one biographer wrote, her "most mystical-sounding ideas stemmed from observation and skepticism, not occult visitations." She merely looked and, in looking, loved. "As you look at these things," McClintock said of her maize genes, "they become part of you. And you forget yourself. The main thing about it is that you forget yourself." How many spiritual teachers have we heard say it—that the point is to forget ourselves so we can merge, in compassion, with the whole?

This same biographer observed that McClintock apparently never had an intimate relationship in her life. I disagree; I think he just didn't recognize her partner. Like a monastic, she spent long hours in communion with her spiritual friend, but in her case, the friend was very much of this Earth: McClintock was intimate with maize.

~

SITTING AND LOOKING at a tree—just looking, intensely looking, until you realize your kinship with it—in some cultures is the equivalent of a school curriculum. Malidoma Somé, a Dagara man from Burkina Faso, tells how he sat with a tree while he was undergoing initiation rites in his village.

I got to know Malidoma more than a dozen years ago when I was called in to help edit his book *The Healing Wisdom of Africa*. He was forty years old by that time, held two doctorates from Western universities, and had been commissioned, with his then-wife, Sobonfu, by the elders of his village to move to North America and teach village values here. Community, ritual, nature—these were the foundations of health and balance in village culture, he wrote, and they were conspicuously absent from life in Western culture. Malidoma and Sobonfu were to teach Westerners how to connect with one another and with the more-than-human forces of nature in order to bring the "hot" Western culture into harmony, for village elders saw that the tempestuous, violence-prone West was about to "burn up" the rest of the Earth through its own imbalance, and becoming more deeply rooted in nature would help to "cool" its aggressions.

Ritual was the path to connecting with nature, wrote Malidoma—ritual conceived of as humans gathering to deepen their relationships with the beings of nature and with the spiritual helpers both seen and unseen. The ritual path would lead to deepened community and more harmonious living with others. Rites of initiation in nature were especially important, for nature is the source of knowing; to know your own destiny, your gifts and purpose in life, you must first grapple with nature's mysteries, and only by forging your own relationship with nature can you remember why you are here and develop the power to release your gifts into the world.

I read Malidoma's manuscript with growing excitement. Here was a mysticism I'd longed for since childhood, without knowing that this was what I'd been missing. But among his people this view of nature was not "mysticism," the prerogative of a strange, wild-eyed few, but rather everyday fact, the knowledge that every person needs in order to live a happy, harmonious adult life.

Malidoma too underwent initiation rites when he returned to his village as a young adult after being educated at a French mission school. During one portion of the tests, Malidoma and the other boys were told to pick a tree, sit with it, and "look at it until its true nature is revealed." Raised on algebra and Kant, Malidoma thought this exercise stupid, and when he sat down in front of a tree, he found himself staring at nothing. The boys around him, who had been raised at home in the village, finished quickly and moved on to their next assignment, while Malidoma sweltered in the hundred-plus-degree heat, gravel digging at his bare backside.

After he had sweated and fumed for hours, getting nowhere, he says that something finally shifted inside him. In the space of a moment he found himself speaking to the tree "as if I had finally discovered that it had a life of its own." Relating now to the tree as another being worthy of respect, he discovered the tree could be a confidant. He began pouring out his discontent to the tree, his anger at having been sent to the mission school, his frustration at not being able to do this assignment.

What happened next changed his life. Suddenly the tree disappeared, and in its place was a green woman, very tall, her veins flowing with emerald light, her color "the expression of immeasurable love." She held out loving arms to him.

A jolt ran through his body, and he stumbled toward her, weeping. He grasped her and felt himself held in return, and into his heart and body flowed a profound sense of acceptance. The sense of exile that had eaten into his spirit during years at the mission school was healed, and in its place was born a sense of belonging. He clung to the green woman tightly, overjoyed at coming home. He held on even as she began to fade from view, and eventually he looked up to discover that he was clinging to the tree.

With that experience, Malidoma says, "my jumpy, doubting mind began to find some rest." Meeting the tree with respect, "as if it had a life of its own," healed the deep rift in his soul. The compassionate person he found in the tree connected him again with the world he had left behind in the village, a world where nature speaks with many voices to those who pay attention. "She had brought me back home."

Malidoma is now a village elder. He still teaches in the West for the better part of the year, and he likes to ask Westerners, "How would we need to change if we granted to a tree the kind of life that we usually reserve for so-called intelligent beings?" Maybe, he suggests, the world of nature is "far more sophisticated than its physical ruggedness reveals"; maybe it is open to communicating if only "we enter nature with a little more openness, and even with a sense of quest."

⌒

TO EXPERIENCE TREES as intelligent beings is to defy centuries of cosmology in the Western world. Yet Malidoma's is but one way of experiencing friendship with other creatures. Gazing deeply with curiosity and respect, as do naturalists or contemplatives, can take us miles toward realizing our kinship with the more-than-human world. And merely sitting a few minutes with a tree, if we let it, will transform us.

I learned to take time with trees because illness forced me to, reducing my life to the simplest possible activities of lying down and looking. Watching the beings of nature seemed a logical extension of watching and listening to my body—no matter that those beings were made of bark and cellulose and sap instead of skin and bones and blood. I didn't see a face in the birch tree who had been my childhood friend; I didn't contact a personality in the bougainvillea. My experiences don't even rank as mystical in the way that term is often defined by scholars of religion. But in the end, it is not the extremes of the extraordinary that lead us home; it is the feeling of kinship that shows us where we belong. When we open ourselves with trust and respect to the creatures around us, the journey home has begun.

When I was in my early thirties and the birch tree visited me to let me know of its passing, I was only beginning to grasp the possibility of connecting with trees as friends. Now, more than twenty years later, I am still just a beginner. I know how fragile any communication can be—even with other humans, even when we speak the same language.

Yet what the intervening years have taught me is how vital such connections with nature are—how renewing for us as humans, and also how necessary if the Earth as we know it is to survive. People who experience a feeling of connection with nature, especially people who grow up feeling connected as children to the nature around them, are less likely to act in the heedless ways that are now destroying our air, water, soil, and environment.

And if we deepen those connections by regarding other creatures as having messages to offer and wisdom to share, are we not treating them as beings of mind and spirit? What if "having a soul," after all, is less about possessing a disembodied core that lives on after death in some other world than it is about contributing to the marvelous complexity of *this* world? What if it means simply taking part in the life force that flows through all earthly veins? If it does, then we needn't be surprised when responding to a tree as if it had a life of its own, as Malidoma found, brings about more than we bargained for. In meeting a tree with respect, we may just be surprised at what—or who—we find.

2

Hooded Oriole

*I pray to the birds because I believe they will carry the message
of my heart upward. I pray to them because I believe in their existence, the way their songs begin and end each day—the invoca-
tions and benedictions of Earth. I pray to the birds because they
remind me of what I love rather than what I fear. And at the end
of my prayers, they teach me how to listen.*

— TERRY TEMPEST WILLIAMS, *REFUGE*

THE SPRING SUNSHINE was warm on my back as I crossed the boulevard and started up the incline to the city rose garden. It was May
1992, and I was well enough now, nearly two and a half years into illness,
to be up for most of every day, and if I moved slowly I could even—oh,
joy!—easily walk several blocks before returning home for a nap.

A few steps from the garden gate, I looked down to see a tiny body
flat on the ground—a dead something or other. I bent closer, sadness
catching in my throat, then saw it twitch. It was a baby bird, all pink-
naked except for bits of gray fuzz. I knew nothing about birds, didn't

want the responsibility of caring for one. Gingerly I touched it—now it really wriggled—then slipped it off the sidewalk under the safety of bushes.

In the garden I could not focus at all on the roses bursting in their first spring radiance or on the walking meditation I usually did there. All I could see was the image of the tiny creature fallen from its nest. I decided that if the bird was still there when I walked past on my way home—if one of the many feral cats hadn't carried it off yet—I would pick it up and care for it for a little while. Almost hoping the bird would be gone, I turned immediately for home.

But there it was, still struggling under the bush. I leaned closer, a sense of horror rising in my chest at something so small and vulnerable smashed on hard cement. And then I picked it up. Such a tiny thing, hardly more than a coin in the palm! Cupping it in my hands, with butterflies in my stomach, I walked home.

⌒

ABOUT THAT TIME, in 1992, Laguna Pueblo author Carol Lee Sanchez was writing, "There are trees and grasses and flowers and birds and ants and bees waiting for you . . . to say hello to them—to call them sister, brother, cousin, or friend. They *are* your relatives; they hear your thoughts as you travel around your town or city." But it would be years before I found her words and even longer before I understood them. Despite encounters with birch tree and redwood and bougainvillea, I didn't yet have words for how human and bird might be related.

Too, I was forgetting something odd that had happened two days earlier—about the hour that the pink-naked being had pecked his way out of his egg.

I had been sitting at my computer recording my dreams in my journal. One puzzling image felt incomplete—a camera stuffed with old film. I reentered the dream with active imagination, a Jungian exercise for allowing the conscious mind to partner with the unconscious instead of dominating it. Closing my eyes, I brought to mind the old camera. What needed to happen? Perhaps a new roll of film? I took the old one out, inserted a new one, and waited. Across my mental screen floated images unusually vivid with life—an image of a green leaf, then a bird with an orangish breast. There was distance between them; the bird had lost its connection to the leafy branch. When I cleared a path from the bird to the branch, the images stopped and I felt a sense of completion. I opened my eyes and wrote the sequence down.

And promptly forgot all about it.

*

IN THE KITCHEN I wondered, "What do baby birds eat?" My husband had no idea, and neither did I.

We gave this one a poor prognosis. A smatter of dried blood on the abdomen—probably internal injuries. At the time I didn't know of any wild birds surviving under human care; I thought that if a wild animal was injured, it was best to let nature take its course. I'd heard too—so many things I didn't know yet—that a mother bird would reject a baby if humans touched it because she could detect the scent. The truth is that songbirds have little use for smell and are not disturbed by a whiff of human touch. The best thing to do for a fallen hatchling is to climb the tree and place it back in the nest so Mom and Dad can resume feeding. The second-best thing is to build a makeshift nest in a nearby tree, and the parents will tend both nests.

But I didn't yet know these things, and it would have made no difference anyway because hooded orioles nest in the stratosphere. Had I looked up from the sidewalk where he fell, I would have seen a brown pendulous basket more than a foot long strung from one impossibly high frond of a palm tree.

Cradling the inch-and-a-half-long body carefully in one hand, I retrieved a ceramic cup, tore up a paper towel to line it, placed the tiny thing in its new nest, and set the cup under my desk lamp to keep it slightly warm. But what to feed the bird? It was Sunday, no experts were on hand, and a Google search was still some years away. So my husband and I took our best guess: bread crumbs soaked in hamburger juice. (I learned later I could have killed the bird with this diet.) Every so often the baby raised its head and gaped, and I shoveled in a morsel of brothy bread from the toothpick of my Swiss army knife.

After a couple of hours the bird started emitting chirps so delicate they were like the squeaks of ancient floorboard under thick carpet. The beak opened, and I obliged. In between feedings I gazed at him like every new parent gazes astonished at a tiny life come so fresh into the world. Every millimeter of him captivated me—the barest fuzz on the top of his head and on each segment of wing; the bulging gray eyeballs under tiny, tight slits; the slightest stumps of charcoal-dark feather shafts on his wingtips; the tiny rear end that put out as fast as I could shovel in to the oversized beak.

One second I despaired I could ever save this creature. The next second he would stir, turn his heavy beak the other direction in the cup, and fall back asleep, and my heart would jump, seeing him persist in life. Whether he lived was not mine to decide. I was there to feed him, make him feel cared for—at least for the few hours until he died—and enjoy him.

So I settled in to a new routine. I stirred to the sound of his "cheep," which grew in intensity over the next hours. If I didn't shovel in the speck of food right away, he cheeped louder. If I heard nothing for an hour, I whistled softly beside the cup. By evening he raised his dinosaur head on its thin wobbly neck at the first tweak of my whistle.

That evening I could hardly take my eyes off him long enough to go to sleep. Who would look after him in the night? I put the cup on an insulated heating pad on the floor and woke up once to check on him. The universe would have to keep him breathing.

～

THE BIRD WAS an oriole, probably male, they told me at a wildlife rehab center later that week when I finally took him in.

I need to make this clear: I didn't do the right thing. Keeping a rescued baby bird at home dramatically lowers its chances of survival. Not to mention, if the bird is a native, keeping it is illegal without a federal permit.

But that week the law seemed the least of my worries. I'd headed to the rose garden to contemplate an excruciating step: after thirteen years of lackluster marriage, my husband and I were considering separation. Long hours of couples' therapy had not shortened the gap between us, yet I was torn: divorce seemed unthinkable. After all, he was a good man, an upright man. Plus, we shared intellectual passions—he a physicist, I a doctoral student in religion—and in the early bloom of our relationship we had dreamed of returning to the alma mater where we'd met to team-teach science and religion.

But as the years passed I'd watched doors close on more intimate possibilities. One night at dinner—several years ago at this point—on the quiet, candlelit patio of a restaurant suspended over San Francisco Bay, he had mentioned a dream from the night before. I was all ears. How did the dream leave him feeling? Did it remind him of anything else in his life? For me a dream was a hint, a clue in the great treasure hunt of living. I could spend hours in the twilight of early morning following each trail of breadcrumbs scattered in the night. But my husband preferred a sunlit faith, a rational faith, and dreams to him were the most irrational part of the mind, little more than detritus from an overactive brain. In response to my questions about his dream, he threw me a hard look and said coldly, "I don't want to talk about it. Now—or ever." I simply stared. An industrial-size door had clanged down between us.

So by the time the baby bird showed up, a dreadful choice loomed, and I was terrified. How would I live? Thirty-five years old, a full-time graduate student, and still so disabled by CFIDS that just to remember the details of life I had to record them in my journal. True, I was recovering, but between struggling to write a dissertation and taking a three-hour nap every afternoon, how could I possibly earn enough money to survive?

To complicate matters—doesn't life often work this way?—my eighty-year-old father had died unexpectedly some weeks earlier, and my emotions were now racing wildly between sadness and relief. To my father, as to many traditional Mennonites, divorce was the unforgivable sin, and I had dreaded bringing him any news of it. My mother would never hear the terrible tidings; her memory had seeped away a few years before like water through cupped hands, the work of a slowly spreading brain tumor, and now she slept out her days and nights in a nursing home. But my father—such energy it would have taken to deal with his devastation! Yet one evening at the end of March he'd bowed his head one last time over dinner, sparing both of us that heartbreak. And in a strange cosmic twist—I felt guilty for appreciating the irony—he would bankroll my divorce: a small sum of money released at his death would help me transition to singleness. It was a financial cushion, or at least the hint of one, and so I began apartment hunting.

༄

ON MONDAY MORNING the bird was still alive, though sluggish. (I learned later that I'd as good as refrigerated him in the night.) Through the phone book I located a wildlife rehabilitation center in the area and called. The woman at the other end sounded horrified at what I had been feeding him and in a clipped voice gave me orders: Clean out its system with a solution of 1 tablespoon sugar and 1 teaspoon salt to 1 quart water. Feed with a dropper until its poop is clear. Then soak 1/2 cup of dog kibble in 1/2 cup of water, add 1 finely diced hardboiled egg, and mash it up together. Feed every 20 minutes. And *please* bring it in.

I followed instructions, cleaning out the bird's system, preparing the mash, and offering it three times an hour whether he asked for it or not. Soon whistling by the cup was no longer needed, for he lifted his heavy head every ten to fifteen minutes and belted out his one syllable. And I turned obediently from my computer, pinched off a tiny glob of food—we'd graduated to tweezers now—and slipped it into his gaping mouth.

But bring the bird in? No way. I was having too much fun.

By Tuesday the plastic-looking feather shafts on his wings were noticeably longer. More fuzz was accumulating all over his body so that he could barely be distinguished from the cotton balls among which he slept. Every so often he'd cheep, I'd shovel in food, he'd sink down again, and I'd turn to my monitor for ten more minutes. At night I adjusted the temperature so he no longer nearly froze.

On Wednesday, since the bird had decided to live, it was time to be grateful for my time with him and surrender him to the wildlife center. Packing the cup carefully into a shoebox, I drove the twenty minutes to the center. I placed my tiny package on the counter, and it was whisked away to a back room for identification. The woman returned holding an empty cup and a paper with an accession number penned in blue. It's an oriole, she said, probably male. We'll try to raise him to adulthood. You can call if you want to check up on him.

That was it, no good-byes.

I walked slowly back to my car, sat down in the driver's seat, and burst into tears. Something about this baby bird, and about giving him up, was shaking me to the core. But what was it?

In a flash I remembered the dream—the bird with an orangish breast, the green leaves, the reuniting.

Inner and outer worlds suddenly merged. I had rescued a baby bird fallen from his nest. The bird turned out to be an oriole, a bird with an orangish breast. The dream had arrived two days before the waking-world bird dropped to the ground. I had made it possible for him to be reintroduced to green leaves.

~

THE NESTLING HAD shown me how well Jung's active imagination exercise worked to harmonize dreaming and daytime lives. But I couldn't go along with Jung's idea of why. I liked his transpersonal theory, the notion that when we dip into dreams we tap wells of knowing that transcend the boundaries of the individual. But I was dismayed at the same time; what Jung called the collective unconscious looked far from collective to me.

Take, for example, the archetypes. For Jung they were built-in templates for thinking and feeling. We inherit them, he said, just by being born and then fill them with the flesh of our individual experiences. Mothers and fathers, for example, may look different around the world, but they are all expressions of the templates, or archetypes, of Mother and Father. And here is where Jung gets controversial. He thought the archetypes are universal. Because our psyches are linked to our brains, and because human brains are more or less the same, he thought the human templates for thinking and feeling are also pretty much the same around the world. So if a man in Europe dreams of, say, Mexican dancers, it is because he is tapping in to this deep well of universal knowledge, an archetype that includes Mexican folk motifs.

I had never been persuaded by his theory. Yes, every human being shares certain experiences, such as being born of a mother and father. But does our similar brain structure really mean that our fantastically varied experiences all point back to a universal well of knowledge? I wanted to respect our differing cultures more than that. Is it possible even to speak of a universal Great Mother, as Jung did, when the actual human experiences of mothers are so drastically different? For example, for traditional Diné (Navajo) people, the home and surrounding lands are the domain of the mother of the family and belong to her in the sense that when she dies they are transferred to her daughter and then her daughter after that, down through the mother's line. In my culture, by contrast, mothers traditionally could not even hold title to the ground they tilled and tended, and until very recently they themselves were the property of their husbands. Between these two cultures, how different not only the perceptions of mothers must be but even the concept of mother itself! With such radically different meanings, is it really possible to say that all instances of mother point back to the same universal archetype? Jung thought so.

To me, Jung's archetypes instead looked European, with a European love of universal mind, an idea that stretched all the way back to Plato and his "forms," those radiant essences of the mental world untouched by the messy world of the senses. Even Jung's way of shaping the question looked Western, for the distance he saw between the supposedly universal essences and everyday flesh-and-blood living appeared to me to be the relic of a long cultural tradition with its roots in the ancient Mediterranean, a tradition that overvalued the mind and undervalued the body. Like Plato, Jung seemed to be grasping for a world unadulterated by matter, a pure world imagined as objective, universal, the same for everyone.

I could not follow him on this one. For one thing, we were separated by too much time. Events that he witnessed in his lifetime, such as the felling of European colonial powers, became understood only in mine. Any chance at a universal narrative, I knew, was now over. Voices from the margins were saying that there never had been a universal narrative in the first place; there were only people in the centers of power passing off their own stories as universal. Feminists had long pointed out that Jung's notions of gender were deeply rooted in his own biography—in the conventions of his time and in his relationships as a man with his mother, wife, and mistress; now I was beginning to reckon with indigenous critiques as well. The privileged, say many indigenous people, have the luxury of thinking their own story is known the world over. Those at the centers of power feel entitled to read the myths of others as variants of their own. Barbara Alice Mann, of Seneca descent, calls Jungian archetypes "another face of colonialism." They do not in the least resemble her people's stories. "It is only the overweening vanity of Eurocentrism that superimposes its own narratives over Native tales this way." Instead of looking for what it might learn from stories so different from its own, Western culture looks to remake others' stories into its own image, "jubilantly announcing to the world that it has discovered authentic 'Native' frameworks and—guess what!—they look *exactly* like the pre-existing metanarratives of Christian Europe." This is a "false universalization," she says, "the *our*-size-fits-all mentality at work."

Just because an image is transpersonal, I suspected, does not mean it's universal.

~

WITHIN DAYS OF relinquishing the baby oriole, my husband and I decided at last to separate, and my apartment hunting intensified. To comfort myself in the following weeks I walked back up the sidewalk toward the rose garden and looked up. There, waving high overhead, was the pendulous oriole basket. Raucous "cheeps," now full-fledged squawks of growing siblings, reached my ears. I smiled and peered at the nest through binoculars, astounded. How in the world had this newborn managed to fall out of such a deep, almost completely enclosed cavity? He must have worked his way up more than a foot of nest wall, a daunting feat for a hatchling. What a lot of effort only to fall, smashed, on hard cement!

I too felt like I was falling. After each foray to yet another apartment for rent, I retreated with buzzing head to my bedroom to nap, convinced in one moment that this move was crazy, convinced in the next I had to go through with it. I was losing every security I had ever known—marriage, family life, the respect of friends who thought our divorce foolish. My mother-in-law, heartbroken, suggested in a letter that I was being selfish. On the heels of my father's death, I was losing her and my father-in-law as well. I spent some part of each day in panicked tears, terrified of facing the world so alone.

During the tumultuous weeks of finding a place to live, untangling the possessions of thirteen years of marriage down to the cards in the recipe box, and wrenching myself away from a house and redwood tree and bougainvillea I loved, the baby oriole receded to the background. I called the wildlife rehab center periodically, recited his accession number, and was told he was doing fine.

The middle of August, some three months later, found me settled alone in a new apartment, tentatively starting a new life. I followed up one last time on the oriole. This time I was told he had been released into the wild at the end of July. I grinned and jumped in the air, fist pumping. He flew!

Only later, after I took up volunteering at a different wildlife rehab center feeding rescued baby birds, would I learn that orioles seldom survive long enough to be released into the wild. Paging through the center's statistics, I would search for "Oriole" and find it near the bottom of the list, chances near zero. Almost all died within twenty-four hours of intake.

~

So how did this creature survive? And how did he find his way into my nebulous dream world two whole days before I found him in the literally concrete outer one? To such a mystery my culture offered hollow-sounding explanations: coincidence, psychological projection, wish fulfillment, fantasy, the stirrings of a repressed id.

And so, for comfort, I found myself trudging back to Jung.

In October 1913 a thirty-eight-year-old Jung was puzzling over his own interior world. A month earlier he had visited his once-beloved mentor, Freud, for the last time. Disagreements between the two had festered for a year, and Jung, hurt by perceived slights, was striking out on his own. But the way was unknown, isolated. Haunted by weird dreams at night and even stranger visions by day, Jung feared he was going insane. Yet how could he help his patients if he refused to do his own inner work? So he recorded the images, marvelous and terrifying alike, in a journal bound in red.

One day he wrote, "I saw a monstrous flood," with sickening yellow waters that engulfed Europe's low countries from the North Sea to the Alps. Thousands drowned in the catastrophe while the peaks of his own Switzerland stretched taller in response, protecting their inhabitants. "Then the whole sea turned to blood. This vision lasted about one hour." Two weeks later the vision recurred, with more blood. Aghast, Jung interpreted the calamity as a private one. "I . . . decided that I was menaced by a psychosis."

Yet a year later Europe was engulfed in a war, blood filling the valleys north to the sea, a war in which Switzerland remained neutral.

Eventually Jung began to notice: the parallels between private visions and public world were inescapable. But he was reluctant to admit that his visions had objective reality; he was a scientist, after all, and the science of the time had no room for relating subject and object. About this time an inner teacher, a recurring figure, suggested in one vision that Jung was treating thoughts as if he himself had generated them when they should be treated instead "like animals in the forest, or people in a room, or birds in the air." The figure added, "If you should see people in a room, you would not think that you had made those people or that you were responsible for them." Could it be that thoughts were companions, having an independent reality?

~

IN RECENT YEARS it has become popular to assert the opposite: "We create our own reality" is a mantra that bolsters thousands. I find the saying crude and often cruel. It belongs, for instance, on a list of things never to say to people with a chronic illness.

"We create our own reality" is also anthropocentric. Who ever speaks of the power of whales, basil plants, and water to create the world? In focusing so exclusively on humans, the notion that the human mind creates the world cuts us off from others whose destinies are interwoven with ours—the billions of species without whose wisdom we would not have come into being and the thousands of billions of individual lives without whose companionship we would cease to exist. "We are human only in contact, and conviviality, with what is not human," reminds David Abram.

It was, finally, the specter of solipsism that led Jung to insist on the link between inner and outer worlds. He set out to show "that the content of psychic experiences are real, and real not only as my personal experiences, but as collective experiences which others also have." If he did not succeed, he knew, utter isolation awaited him—not merely exile from his scientific profession but also a fate equivalent to madness, the inability to link his own perceptions to those of others.

Following the narrative line of his culture, with its love of universal, objective truth, Jung settled on his theory of archetypes. The archetypes were universal and therefore must have scientific standing. He had finally proved the independent reality of interior, visionary images. Isolation was banished at last.

What he could not see were the limits of his theory. He had charted new intellectual territory, it is true. By asserting an objective reality for the

interior images, he blazed a path that a society steeped in rational material-
ism has yet to follow. He began to heal the age-old split in Western thinking between subject and object, self and other. But he did so at a price: for
him the interior structures of thinking, to be authoritative, had to be not
merely transpersonal—having a reality beyond the caprice of fantasy—but
also universal, the same (or similar) the world around. In this way he was
not so different after all from his scientific colleagues: like them, he measured truth by its distance from unique, subjective experience.

It would be left for others to see how deeply tied his archetypes remained to his own personal and cultural ground. The archetypes encoded
his own perceptions after all. Intimate relationships with significant others
were extracted from the rich but messy stew of individual life and painted
bloodless on a Platonic screen. Stories were condensed to principles, relationships abstracted into archetypes.

We needn't be surprised if the archetypes don't fly well across colonial
borders. Only stories, firmly fleshed, can grow such wings.

~

AND SO WE return to stories, as Jung too returned decades later, offering at age seventy-six his theory of synchronicity—or seeing with
inner sight something that proceeds to come true in the outer world.
Synchronicity is often described as a "meaningful coincidence," a feeling
of connection between two events that have no cause-and-effect relationship between them.

Jung told the defining story. A patient of his, emotionally rigid, has
retreated into a brittle rationality. In their psychoanalytic sessions together
they explore her dreams but make little progress; her inflexibility is winning. Then one day she describes a dream: she has been given a gift, an
expensive piece of gold jewelry shaped like a scarab beetle. Even as she is
speaking there comes a *tap-tap* on the window, and when Jung opens it,

in flies a large beetle of a green-gold color. Jung catches it and hands it to her: "Here is your scarab." Outer and inner worlds mesh. The patient is catapulted into a new way of living.

Eventually Jung came to believe that synchronous events said something not just about the human mind's ability to find meaning but about the world itself. *Unus mundus*, "one world": inner sight and outer world are the same; the psyche is one with the physical world.

Today we might say it even more plainly: the psyche is one with *nature*.

The reality that we plumb when we reach into our own unconscious is the world of leafbuds and honeybees, mollusks and meadowgrass, coots and humans. But not, as the younger Jung thought, some universal, abstract reality stripped of personality and place. In dreams and visions we may indeed touch places that we don't know—or don't yet know—but just as often, the symbols we encounter may be crafted from the nature at our doorstep: the trees that meet our eyes as we walk to the bus stop, the neighborhood in which we're immersed and from whose wellsprings we draw our creative juice. These are our relations, both human and more-than-human. Why wouldn't our neighbors and kin want to be in touch with us? Or, the same thing, why wouldn't our unconscious minds know their own surroundings? Is there any reason to think that just because the psyche is larger than conscious awareness, it won't also start with the place where we are?

~

IT IS FITTING, as we try to remember how to love our companions on Earth, that so many stories of synchronicity involve animals. I think of an April-gray afternoon a few years ago, soon after I moved to Boulder, Colorado. Tired of being cooped up through an icy spring, I piled on layers of clothing and headed out to bird-watch at the ponds outside of town. On a boardwalk meandering across winter-brown reeds, I approached a white-haired man with binoculars. "Who's out today?" I asked.

"Oh, I don't know yet," he responded. "I haven't been here long. But," he added, "the yellow-headed blackbirds are due back any day now." I brightened. Maybe I'd catch my first-ever glimpse of one.

We compared notes for a moment then turned back to the wetland. Not more than a minute passed before a bird the size of a robin flew in and perched on still-naked branches thirty or so feet ahead of us. The gathering twilight obscured details, so I reached for my binoculars—and was startled to see a flaming yellow head above a glistening black body. A male yellow-headed blackbird, just in from his southerly migration! I turned excitedly toward the other birder, who at the same moment turned toward me.

"What did I tell you?" he exulted. "And just as we were talking about him!"

One world: a synchronicity shows us just how close inner and outer worlds like to sit, like new lovers mesmerized in each other's presence. Subject and object are not separate; a synchronicity is their love story.

And there was my experience this very morning. Before dawn I dreamed for the first time ever about a bear. A cinnamon-colored bear simply sat upright and stared at me gravely. A second bear seemed to lurk in the shadowy background.

I woke up and turned the dream over in my mind, wondering what it meant, then got up and headed to breakfast, picking up the local paper. A large headline across the front page read: "South Boulder Bear Killed." A female bear had wandered into local backyards several weeks before and had been tagged and relocated to a different county. Yesterday she had found her way back to town and was spotted too close to an elementary school. Two trips to town spell doom for bears, and the Colorado Division of Wildlife had felled her near the school. She was believed, said the paper, to be the mother of a male cub who had died accidentally the previous month.

I couldn't help but think she had returned because she was grieving her cub. What else to do but revisit the places she had shared with him? And of course humans might hear the tale in dreams. Why wouldn't the bears also publish their story?

⌒

IF THE PSYCHE is one with nature, then perception is tied to place. Knowing comes with biography and landscape attached. Insight wears a time-date stamp. The genius of our own inner knowing is not that it is universal but that it is particular: it shows what is true in this place at this time. Or, to turn the equation around, the genius of nature is not that it reveals universal knowledge but instead that it reveals the knowledge that each person needs right here, right now. Nature's brilliance lies in meeting each of us where we are. How could it not? Nature *is* us—blackbird and tarantula; bear, arugula, and human.

If the psyche is one with nature, then nature is relational. People of all kinds—human, tree, animal, wind, lake, and dream people—are waiting for us to acknowledge them, to say hello as we pass by. "The universe is personal," wrote theologian Vine Deloria Jr. of the Standing Rock Dakota nation, "and, therefore, must be approached in a personal manner."

To approach the world in a personal manner is to live in the particular rather than the abstract. It is to pay attention to place—where place simply means, said Deloria, "the relationship of things to each other." It is to reaffirm connection as essential to knowing and especially to healing our ecologically broken world. Instead of regarding the particular as a burden to be shed in the search for a truth imagined as independent of place and time, we will look to the anomalies of nesting orioles or beetle-chewed pines to help us navigate a world pushed to the brink of destruction by our disregard for place. We will learn to nurture relationships.

~

ON MY LAST night with the baby oriole, I hung many minutes over the edge of the mattress to gaze at this life-in-miniature on the floor beside me, his abdomen pumping quick and lively in sleep. I did not see the parallels between us: losing our familiar places, flung out into the world to find— what would it be?—feral cats or friendly faces. I didn't yet hear his message about how closely inner and outer worlds like to sit, did not yet remember the dreaming that preceded him. All I knew was that in tending the bird for three days I had experienced a bone-deep delight in his tiny tenacious life.

An immense gratitude welled up in me—as if finding and caring for this creature was the most joyful thing that could be done, ever. I was flooded with awareness—how precious this tiniest being is to the life of the world! A feeling of wonder splashed over me. Some dry part of me, buried unnoticed in despair, freshened again.

In the months to come, I would need this fragile stirring of hope.

3

Wild Orphans

To go in the dark with a light is to know the light.
To know the dark, go dark. Go without sight
And find that the dark, too, blooms and sings
And is traveled by dark feet and dark wings.

— Wendell Berry, "To Know the Dark"

As it happened in that year of divorce and death of my father, the momentous losses were not yet finished; my mother too died at the end of December, the brain tumor snuffing out the last of her life systems.

Yet for the six months prior to her death, I felt rejuvenated. I was thirty-five and finally living on my own. As soon as I moved out from my marriage, all the panic and dread I'd experienced in the months leading up to it melted away, dispelled like night phantoms in the clear light of dawn. The step that for years I'd been frightened of taking now lay behind me, and I exulted in my new freedom.

Not that all of life was rosy; I'd just experienced two jarring losses—of father and marriage—and I understood the inner rhythms of feeling. I took time to be silent, to cry, to feel the enormity of the changes I had just made.

69

But crying worked its magic; each time I drained the rain barrel of tears, it stood empty for a while, ready again for sunnier weather.

My body too was healing. Although I still took a long nap every afternoon, I felt nearly whole again. My new apartment sat only a block from the sparkling, blue-gray jewel of Lake Merritt in downtown Oakland, and I walked often to its peaceful gardens to sit on a bench and listen to bees or watch hummingbirds dive-bomb in breathtaking aerobatics. Before falling ill, I'd regularly walked around the lake, marveling as gray pelicans plummeted from the sky, bill first, to catch fish, but by now three years had passed since I'd been able to complete that circle. One day in August I started slowly around the lake—and discovered that if I took it easy I could keep going. Two and a half hours later, with energy to spare, I strode up the walk to my apartment. Three whole miles! I jumped up and down, jubilant. I began alternating afternoon naps with lakeside walks. It was a turning point; from here on, recovery meant figuring out how to live the waking hours more than taking time for sleep, and over the next year and a half the naps would decrease until they finally dwindled to nothing.

A few weeks later, on Labor Day, I drove up the coastal ridge hemming the eastern edge of town to explore the regional parks at the crest of the Oakland hills. In Joaquin Miller Park I found redwoods by the hundreds, and I strolled silently in the shadowy, cool forest, enjoying that special hush found only in their groves. I sat beside the trail, leaning up against a young redwood, and marveled at the quiet. A low thrum vibrated at the edge of the stillness, muffled evidence of the city lying just below this hill. Across the trail, a small bird—a brown creeper, I later learned—hopped headfirst down a redwood trunk and paused at the moss-glazed ruins of an old drinking fountain to sip from the puddle in its concrete bowl.

On later visits I found fairy rings thirty feet across inscribing circles where the old *Sequoia sempervirens* giants had stood before being ripped away in the late 1800s to line posh homes and buildings. Young trunks—clones of the parent, carrying the exact same genes—marked the perimeters of the aged trunks, and within those enormous circles the profound stillness of redwoods shifted to quiet joy. Hushed reverence danced, alive, in the memory of giants.

I was nearly well in other ways also, for that autumn I fell in love. I'd started dating as soon as I was single, but after a few false starts I'd discovered that the boyfriend of my dreams lived under the same roof, the other renter in this stately old triplex. We got to know each other slowly, eating late-night snacks or meeting in morning sunshine in the front yard to admire enormous orb weaver webs bejeweled with dew. By Thanksgiving we were riding his motorcycle up through the greening hills of Marin County, and by Christmas he was taking me skiing at Tahoe. He taught me to play—the kind of play I'd longed for but didn't know adults actually *did*—and as my strength returned, I dug in deep.

But the day after we returned from our first ski trip, at Christmas of 1992, my mother died, and I began a long slide into despair. I still found plenty to celebrate, like learning to ride a mountain bike in the summer of 1993 and discovering I was strong enough now to ride five miles, even uphill. The April day in 1994 when I finally bought my own bike marked the official end of illness, more than four years after the bout of flu that signaled its start. I now was so healthy I needed regular, vigorous exercise.

But while I had been growing stronger physically, my emotional health had slipped. Raw grief just after my mother's death slid over subsequent months into prolonged and wrenching sadness. I didn't know how to shake the dogged sense of loss. Three primary relationships—with mother, father, husband—were now over for good. I expected to feel bad for a time, but as hard as those relationships had been—three of the toughest of my life—why couldn't I feel even the tiniest bit of relief?

I was taken by surprise when all the tools that had served me well up to now did not resolve the grief. Feeling the feelings, telling the stories of abandonment and neglect, processing them in journals and with therapists and healers—I'd been doing all these things for years, and yet now, after being freed from these three thorny relationships, I felt worse. It only deepened the misery that my boyfriend seemed to be slipping inexorably away.

When, a year and a half after my mother died, the boyfriend finally left, my despair knew no bounds. I was newly single again, this time against my will, and my desolation slid seamlessly into fury. The wrath, of course, was futile; I knew it better than anyone, which only ratcheted up its volume. The best I can say of that period of time is that I feel some relief knowing I will not have to endure it again, for those four losses—father, mother, marriage, lover—cannot constellate twice. And I am forever grateful to the few friends who patiently loved me and kept listening to me through those bleak and angry months.

I was now so wracked with loss that normal conversation took great effort. I avoided small talk and withdrew into myself. Needing to vacate the space forever linked to my former boyfriend's bright blue eyes and cheerful grin, I moved to an in-law apartment in the Oakland hills. It sat on a quiet, dead-end street with views of the San Francisco Bay. The apartment offered sunny mornings and quiet days, and it was in walking distance of the regional parks, where I longed to be healed by the magical calm of the redwoods.

But calm remained beyond reach. Instead, fury seethed. Days of fierce, impotent frustration alternated with nights of grinding insomnia. I was cut loose, adrift, floating in the blankness of outer space, as if in a darkness stretching outward in all directions. Panicked, I tried to meditate, but the disorientation was too profound. Watching my breath brought no comfort in a body agitated by grief. I would sit on my bed at three in the morning, crying enraged tears and pounding the bed with both fists.

I think now—and thought so then, though it was no help at the time—that the deeper problem was isolation. I had one or two friends in town, whom I saw every few weeks, another in Los Angeles three hundred miles away. To have loved ones near on a daily basis would have made an enormous difference—friends to witness and keep watch and help absorb the shock of change. Alone I was no match, as none of us are, for the searing winds of loss. But years of first illness and then divorce and deaths had eroded my support network; I lived alone in an urban area as a student whose friends kept graduating and moving away; and I dwelled in fury and pain—not a good state in which to make new friends.

I kept myself busy with editing work since trying to collect my thoughts enough to write a dissertation was often impossible. In a rich irony, I was given to copyedit during that time a small volume of poems for insomniacs called *Prayers at 3 A.M.* as well as a book that became wildly popular, Coleman Barks's translations of the thirteenth-century Muslim mystic, *The Essential Rumi.* I completed both of them in the bleak hours before dawn.

↶

I HAD ANOTHER problem too. I had no faith. I couldn't accept death.

I'm speaking now of death in its largest sense—all the things we have no control over. A face of death greets us in aging and illness, in childhood traumas, in the loss of a partner or an accident that leaves us impaired. If dying at the end of life is inevitable, then just as certain are the daily reminders that loss accompanies us in every moment, a bass drone rumbling just below the surface of everyday life.

For me the drone of dying took the form of broken, long-term relationships that would not be healed no matter how hard I worked at them. In this sense my life up to this point had been filled with a great deal of dying, for these relationships were the first and most intimate one can have—with mother and father, then with a husband and later a lover through whom I sought to clean up the earlier messes. I'd been in enough therapy to know that adult relationships follow logically from patterns learned with parents.

I could even pinpoint the problem: a mother who was emotionally and physically ill and a father who stood by helplessly, wringing his hands. But this knowledge brought no relief from pain. I was alone, I was unmoored, and, in the habit of the grieving—we tend to take our losses personally—I felt cheated in a cosmic way by being handed a history of family dysfunction. I desperately wanted the past to be different than it was.

During a visit to my mother in the nursing home a few years earlier, just before the brain tumor that killed her grew so large that she could no longer recognize me, she had reminded me how important it was to "work with your hands," the only possible way she, a former farmer and nurse, could imagine to make a living. Proudly that day in her wheelchair she had asked, "Do you remember how I always called you to help with dinner?" Of course I remembered—half a lifetime's worth of being jolted out of a reverie with a book by her insistent voice calling me to the kitchen to peel potatoes or take the meat out of the freezer or perform some other mindless task. Was she telling me now that she'd interrupted me those thousands of times *on purpose?* I stared at her, unbelieving. "You can't spend your life in books," she said firmly, her tumor-blinded eyes gazing somewhere toward my face.

I teetered between outrage and amusement. I was already a doctoral student and book editor. I studied books, corrected books, taught books, made my living by reading books. What I'd long suspected was true after all: my mother had never been able to see me, not even before her eyesight was gone.

Now she stalked my shadowy dreams, rising from her deathbed night after night, suddenly aware and smiling and able to see me clearly. Morning after morning I awoke to find this mother a mirage. Daily I lost the mother I wished I'd had, just by waking up.

ᕲ

IF MY MOTHER felt she had to squelch my dreaming, it was because she had first squelched hers. My parents grew up between the world wars, their youthful hopes stunted not only by the Great Depression but also by the rigid beliefs of Mennonites in our corner of Ohio during those same years. Tyrannical bishops in the 1930s decreed the details of church members' lives down to the length of the sleeves on women's dresses (below the elbow, of course). Deviating from church-approved standards meant losing not only your community but perhaps your family and your food as well, for the local church at that time shunned its wayward members, not even speaking to or eating meals with those who disobeyed, not even when they were members of their own families.

The bishop system, the cruelty of shunning, and the Depression all had run their courses by the end of World War II, but their deadly work was internalized by a generation who mistrusted desire and whose youthful dreams, crushed by hard physical work, crystallized into a stoic and severe practicality. Their despair went unverbalized, but you could hear its echo in a grumbling tone of voice or repeated pent-up sighs: life was hard, dreams must die, and parents had to break this terrible news to children so they wouldn't be disappointed in the world they found when they grew up.

My mother showed more alarming symptoms. During my childhood she suffered chronic pain and was medicated, as unhappy women in the fifties and early sixties were, on a cornucopia of prescription drugs beginning with Valium. From time to time her frustrations erupted in fits of rage similar to those of alcoholics. When she was "sober," she was a sweet, timid woman who blushed easily and wanted above all to do what was right. But when she was angry, she loudly berated whoever was nearby for hours or days on end, around the clock, her rage finally fizzling after several days into fits of remorse and apologies. These binges happened more and more often until, by the time I was in high school, they were taking place more than once a month. I spent many nights plugging wet toilet paper in my ears, as my older brother had shown me, so I could catch a few hours of sleep.

Now that she was gone, I was inconsolable. The mother I might have had was lost forever.

I came undone. Chronic illness, by comparison, had been a piece of cake; it had ebbed over time. But no acupuncture treatment or daily meditation was going to bring back the dead, let alone turn them into the people I so desperately wished they had been. The past was unusable, the foundation it should have provided nonexistent. No future could possibly be built on such a heap of nothing.

A wise friend at the time advised me gently, "The cup has to empty before it can refill." But at that time I had no idea a cup *could* refill. Unhappy relationships, one after another, were mounting evidence that cups tended only to drain. Once the cup was dry, all hope was lost.

And so I sank into depression, convinced at some bone-deep core that loss and dying are the whole story. If mania is the feeling that one will live forever, then despair is the feeling that dying, loss, and grief are all there is.

෴

IT SEEMS ABSURD, this focus on one half of the story. One only has to look around to see that spring follows winter, birth takes place just as often as death, and a season of renewal usually follows one of dryness or depletion. But that's just the point: one has to look around—at nature. And if nature is not a part of everyday belief, as it was not part of the Christianity in which I was raised, it is almost impossible to know in one's body—in the heart or especially the belly—that cycles of beginning and ending perform in tandem, leapfrogging as surely as spring and fall. Stories do end, death is real, and loss burns fiery every time. But each story ended also launches a new tale.

I had not lived long enough to learn this in my body, and no religion I had been part of had helped me remember it. Nature was missing even from the religion of the farmers among whom I was raised. This too seems absurd, but it is true. The Mennonites in my farm-studded area, mirroring American Protestantism in general, skipped right past nature to worship a God beyond this world. It was the job of the faithful to remember how different God is from all that we can see or touch, for to lose sight of that distance was to fall into idolatry.

Three times every day we reinforced that immense cleft, closing our eyes and bowing our heads before every meal and thanking an invisible God for providing the food. Absent from any of our minds were the animals or plants who had given their bodies and lives to sustain us. The vegetables and meats on the table were only impersonal "foods" provided by the unseen heavenly deity.

About the time my parents died, I fantasized with horror that I might be called on to say grace at one of the family gatherings—after all, I had a seminary education—and I figured I would have to at least mention the animals, maybe even thank them too. I also knew once I did that, my relatives would be the horrified ones. But as usual, the family grace was said by the man of the house, and for once that was fine with me.

Yet if God was the source of the food, the farmers of the Midwest during my growing-up years were relying on science too. At the urging of agriculture secretary Earl Butz in the sixties, they began pouring tons of industrial herbicides and pesticides onto their fields, and when their yields increased, they gave science more than God the credit. Science was visible in these rural counties while God was not; "he" was tucked away far above Earth in a heaven that rarely intersected with economic worries like crop yields. To grow our food we of course relied, as did all growing things, on the providence of God to send rains at the right time, but God, we all knew, was far removed from soil and grass and trees, and digging in the dirt had little to do with "divine" things. For the Mennonites of our region, as for most evangelical Protestants of the time, "being saved" meant making sure that our souls rejoined God in that sky-heaven after death.

Not that all the attitudes I absorbed about nature were negative. My father marked the seasons with birds. When late-winter snow still blanketed the ground, he counted up the robins he had spotted on his drive to and from work; he was always the first to glimpse an early red-breasted migrant. In summer he crafted small birdhouses to attract wrens and purple martins and was gleeful when they took him up on his offer. Yet the saddest day of the year for him fell at the height of bursting June, the summer solstice, because now the days would only grow shorter. When the robins departed in the fall, he was disconsolate.

From my parents, who were farmers for much of their lives, I learned that humans live by and large in an adversarial relationship with nature. The natural world may be the source of livelihood, but any living taken from it is wrested only through great effort from the ground. One might admire the rows of canned peaches lined up in the cupboard or the green beans filling the freezer, but the joy to be found in them has more to do with being pleased by the fruits of your labors than with taking delight in the ever-blooming, ever-bearing generosity of Earth.

Now, in my frantic despair, I sensed that hope, if it could be found at all, might be found in some way in nature. I decided to spend one day a week—a sabbath—outdoors. At first I did it doggedly, hiking in parks or sitting under a tree or walking along a beach and listening to the ocean's roar. I say doggedly because, at least at first, it did not noticeably help. Despair does not budge just because flowers are blooming. Still, week after week, I kept at it. At least I was getting out of the house.

I remembered too that poetry could provide solace. I'd been drawn some years before to Mary Oliver and Wendell Berry and to the mystics from religions around the world: Rumi, Rilke, Lalleshwari, Tagore,

St. John of the Cross. Now, in my desolation, I searched out the voice of Makeda, Queen of Sheba (ca. 1000 BCE), in lines written by poet Jane Hirshfield, and hoped that one day her words might describe me:

I fell
because of wisdom,
but was not destroyed.

I clung to poetry the way the devout cling to scripture.

My closest friend in town, Mimi, nourished my growing love of poetry. Mimi and I had met over books; she worked at the publishing house that provided most of my contract jobs. Some years before, she had sent me a little manuscript of sentimental spirituality to work on, and we'd become instant friends when we discovered, to our delight, that each of us despised it. Now, whenever my despair deteriorated to angry depression, Mimi began calling me up and reading poems over the phone. I would return home from a walk in the woods or a therapy session and find Mary Oliver or Hafiz or Robert Hass waiting for me on the answering machine. Mimi picked nature poems with simple images of loss followed by equally simple hope; to this day I hear Howard Moss's "The Pruned Tree" in her voice:

As a torn paper might seal up its side,
Or a streak of water stitch itself to silk
And disappear, my wound has been my healing,
And I am made more beautiful by losses.

Was such seamless mending of the soul even possible? Would I ever feel new life moving so deeply within that I could say, like the tree, "Now, I am stirring like a seed in China"?

But the day I came home to Wendell Berry's "Song in a Year of Catastrophe," I thought the tears might never stop. The poet, after wrenching losses, is asked to let go even more deeply. Afraid, he resists losing all, but he hears the persistent call to live even closer to the earth, to let himself be covered by earth. "You mean death, then?" he asks. The answer is, "Yes . . . Die / into what the earth requires of you."

I recognized that relentless march of loss, and also the fear of it. I wanted to think that my losses too belonged to some natural order—above all, I wanted to trust that order—but I could not. Of course I knew that death ends every life, but it was not death itself or the dying of my parents or my marriage that seemed to be causing so much pain; that's why it was so mysterious. The poem sent me pawing for answers in dark closets of despair—then caught me unawares with its final lines:

I let go all holds then, and sank
like a hopeless swimmer into the earth,
and at last came fully into the ease
and the joy of that place,
all my lost ones returning.

Could it be that my lost ones too would return someday? I dared not hope for it. I could not trust it. Yet the words washed over me like the first rain after summer's dryness. I wept and wept, just for the possibility.

৵

YOU'D THINK I would have had the tools to meet this dark night of grief, to stave it off through sheer numbers of spiritual practices. In addition to keeping a dream journal, for years I had kept two other journals— one by hand so I could scribble across a page, the other on computer for pondering and reflecting. I had practiced walking meditation—following my breath, quieting my mind, watching my body and the world around me in a peaceful way. I had attended therapy groups, had been through years of couples' therapy, and every morning and evening had practiced relaxation exercises. The spiritual practices had served me well through years of illness and divorce. What I hadn't realized was that they would not be enough.

The dark night of the soul wouldn't be called that if all the carefully garnered skills did not at some point give way. The night isn't dark until you cannot see where you are going, until all the lights you carefully packed for the journey fade away and dim to nothing. "To go in the dark with a light is to know the light," writes Wendell Berry; "To know the dark, go dark."

And though meditation and therapy and inner work had prepared me for the journey and guided me through new waters, by themselves they could not ferry me to the other side of grief. They were things I practiced alone, in the privacy of my bedroom or the forest or sandy beach, and they did not lead to community with others. Of course, the Bay Area held Vedantic and Buddhist communities and Eastern teachers of every stripe, but though I looked and listened and sampled, I never found either a teacher or a group that called strongly to me. The groups I visited were either too old or too small or too quiet or too noisy, and, although I didn't see this at the time, they left me restless for a deeper reason: they often pointed away from nature, away from the source that was fast becoming my deepest solace. I might sit among worshippers in a decorated room, eager to leave so I could walk on the beach. In the middle of a dharma talk I would find myself wishing instead for the sound of leaves rustling.

And so, lacking a spiritual community, I lacked its tribal connections. Though I tried in every way I knew to find community, it remained elusive. Ironically, although in the worldview I clung to, separateness is a dream— nondualism says we are all parts of one another, all made of the same stuff, closer to one another than the heart is to its own beating—when it came to daily life I had never before (and have not since) been so alone.

Blindly, I continued doing things that were good for me: spending time in nature, reading poetry, taking ceramics classes, mountain biking in the regional parks, beginning therapy again. My new therapist, a woman a year or two younger than me with kind, deep-brown eyes, almost immediately recommended an antidepressant, and I added that too. The medication did not quell the seething depression, but it did collect a bit of ground under my feet, and I was finally able take a first hesitant step toward the future.

In the midst of my bleak and desolate fury, I remembered the simple, straightforward needs of the baby oriole. I longed for connections that required no words. I longed to be of use. I needed the company of other orphans.

Three springs after the baby oriole, I called a small wildlife rehab center about twenty-five minutes away. Yes, they needed volunteers to feed baby birds. Yes, they would train me.

᠈

"NONE OF THE 'world' religions is an earth religion," says Ronald Grimes, a Canadian professor of religion. None of the "great" religions is tied to a place; what makes them "world" religions is that they tore themselves loose from their birthplaces in order to appeal to—and often to conquer—people in other places. The world religions do have holy centers, but their axes mundi are holy not because of looming mountains or flowing rivers but because of the memory of events that happened there to human beings—a crucifixion, an ancient promise, an awakening after a long night of struggle.

Grimes suggests that in this time of ecological crisis, the world religions need to turn their attention to the Earth: "With rites we have served gods; now, with rites let us serve the ground, the air and water, the frogs and rutabaga."

Thankfully, religious people around the world are responding. And though their religions may not begin with the Earth, each tradition provides tools that people of faith are bending to suit this new purpose.

I think of monks in Thailand mobilizing to halt deforestation. Prior to World War II, nearly 75 percent of Thailand was forested; now, only 15 to 20 percent of it is. So much loss of woodland over so few decades means that rivers are silted and fish are dying from lack of water and spawning grounds, and birds have abandoned a country of few trees. Since the 1990s, however, groups of monks have gathered to protect remaining stands of trees. They use a very Buddhist tool, the ritual of ordination. They ordain trees as monks. Local people along with the monks drape the trees in the saffron robes of monastics. Because monks are highly revered in a Buddhist country, the golden robes deter loggers.

I think too of Christians organizing to stop mountaintop removal in Appalachia. Coal mining in recent decades has consisted of blasting the tops off mountains to reach those inaccessible seams of black, then burying hundreds of miles of streams with the cubic tons of waste pushed from the site. But as the destruction continues, Christians are organizing prayer vigils. They host tours to show influential people this ugly source of our national energy. They engage in civil disobedience at the Massey coal company's headquarters in West Virginia to protest the company's actions. "The earth is the Lord's and the fullness thereof," they say, quoting Psalm 24:1, "the world and they that dwell therein." The Reverend Allen Johnson, a guy with a wide-open grin and a close-trimmed white beard, helps organize the network called Christians for the Mountains. "We believe that God made this planet," he says, "that God loves the earth, God loves creation, God loves humanity, and that even though God gives us freedom to spin our destiny, God doesn't want it to be trashed." God has a covenant with those who live in the future too, he says. "How can we have a covenant with future generations when we destroy land permanently?"

⁓

THOUGH I DIDN'T put it into words at the time, I too was reaching for some way to serve the Earth, but my reasons had little to do with grand causes or religious responsibilities. I just wanted to feel better. It was the spring of 1995, and in the three years since the deaths of my parents and the end of my marriage, two other close friendships had ended badly, and my new therapist was about to get married and go on a yearlong, around-the-world honeymoon. (How *could* she?) I was desperate for friends.

Raw with loss, I showed up for my first shift at the wildlife center at 4:00 PM on a Monday. I was to feed baby birds until 8:00. I'd requested this time slot because the center would be closed for most of it, the front intake room empty of visitors, and I wouldn't have to put forth monumental effort to speak with strangers.

Walking into the clinic, I was instantly overwhelmed by smells: faintly acrid disinfectant over a musky something—heavy, close-bodied, like the back rooms of a vet hospital yet different. The cages of all the animals had been cleaned that morning, the hospital was spotless, but the smell alerted me from the start: the ones whose company I was about to keep were unfamiliar, wild.

I dug in to the work. I fed baby birds sheltered in white plastic incubators, tricking them into gaping by blowing gentle puffs of air across their heads, like the soft whoosh of parent wings arriving at a nest, which automatically opens the babies' beaks. I marveled at the tiniest baby hummingbirds brought in with their nest, cut down accidentally by someone pruning a hedge. Hummer babies always come in twos, squeezed into a nest the width of a quarter; the interior is soft with down and its firm, smooth exterior often decorated. Around this one tiny aqua paint chips had been woven in and secured with spiderweb threads.

I stuffed soaked-kibble-and-egg mash down the gullet of juvie ravens and was gratified by the loud, gobbling gurgle that ended each swallow. I learned to force-feed astonished young mourning doves with liquefied kibble and banana, threading the tube feeder down their throats and into their crops with one hand while calming them with a gentle, steady pressure from the other hand around their wings.

The force-feeding was necessary, I learned, because mourning doves feed their young differently from other birds. Rather than dropping food into the beaks of babies, doves secrete a sort of milk from the lining of their crops, a meal that both Mom and Dad are stimulated to produce when a baby puts its beak inside theirs. Baby doves will not gape unless their beak is enclosed by the parent's, and good luck to a rehabber trying to replicate *that*. (I managed to make a juvie dove gape exactly once by grasping its beak for a moment between my thumb and forefinger.) Doves are unusually upset in captivity and always too frightened to eat willingly. Their cages must be draped at all times and the room darkened when opening the cage doors, or young doves will fly frantically about. When they did escape, I learned to catch them too, clambering up on chairs and counters to grasp them lightly out of midair, hands cupped around their bodies to still and protect the fragile wings.

At the wildlife center I learned to handle hawks and owls, but one bird sighting was a special treat. Just a few weeks after I started working, the staff member on duty led me out at dusk across the crunching oak-leaf litter to a small building at the back of the compound. She unlocked the outer door, then the inner one, latching both carefully behind us, and, placing her index finger firmly to her lips, lifted one corner of a cage drape. There was a golden eagle, its ruffed neck dark golden brown and its fierce eyes staring at us. Awe comes naturally in the presence of such a being.

At the other end of the size spectrum were the bushtits, tiny gray-brown birds whose eyes are but pinpoints of black fire and whose minuscule beaks, much tinier than those of other songbird babies, can wrap themselves around mealworms twice their diameter. I always wondered if bushtits were somehow unhooking their tiny jaws to suck in the fat, squiggling worms we offered with tweezers. Older bushtit babies were treated to freshly cut branches of shrubbery we gathered for the aphids on their leaves. The bushtits would move methodically from leaf to leaf, devouring the insects.

Feeding the birds calmed me, at least for the half day a week I was there. I took refuge in the busyness. Stove timers dinged every fifteen minutes (to feed the hummingbirds) or thirty minutes (for the baby songbirds) or on the hour (for the juvenile jays and robins), and even before one round of feeding was done it was time to begin the next. In between, I had to squeeze in the meds: pink penicillin in measured droppers for birds listless with infection or brought in with puncture wounds from a cat. As twilight gathered, the feedings wound down and the birds settled into sleep, but then it was time to sterilize all the droppers and syringes, write up reports, sweep and mop the floor, and lock up for the night.

I did watch birds die—a crow with a broken wing too complex to be set, or new-hatched, naked baby birds who could not be roused from their torpor—but most of my time was spent rushing from one cage of very much alive young birds to the next. I had no time to dwell on my sadness and fury, and while they didn't exactly lift, they did recede, if only for a few hours, and for even that brief respite I was grateful. As the weeks passed, my emotional skin continued to seem thinner than the wrinkled tissue-paper flesh of the desiccated baby birds I was trying to revive with fluids. Yet I noticed, in a small bit of grace, that the very quiet I had retreated to in my depression seemed to soothe them.

In return, their simple needs soothed me. The juvenile jays and robins and insatiable house finches screeched and chirped nonstop for food, filling the two small hospital rooms with a raucous din. But they didn't ask me to talk, they didn't wonder why I was pale or quiet, and they didn't care about my losses. In the face of their simple, matter-of-fact needs, I experienced a simple, quiet relief. They wanted only food and a bit of comfort, and I gave them what I could. To minimize their trauma, I learned to feed them quickly and handle them as little as possible.

I did always try to give them a bit of the hope I couldn't grasp for myself. Each time I restrained a dove to tube-feed it, I telegraphed pictures of oak and bay trees and clear blue sky and a feeling I could only imagine—of using wings to flutter from the ground to branches high above. I knew their captivity would end: soon they would learn to feed themselves with seed, and shortly after, they would be released back to trees and sky. *Don't give up*, I tried to convey; *this darkened room is not the end of your story. Soon you will be flying free.*

4

Red Foxes

When you can see each leaf as a separate thing, you can see the tree.

—YUROK MAN

I SQUATTED ON THE floor, waiting for Rudy to finish pacing. At last he had slipped inside the safety of his night pen, and I had dropped the gate to lock him in. Now his food bowl awaited. But would he approach it with me crouching only two feet away?

The red fox and I had performed our usual evening dance. Me: stand just inside his night pen with my hand on the long rope leading to the gate. Rudy: race around the outside enclosure, dart into then out of the night pen. Me: stand stock-still. Rudy: creep up to the doorway, peer in. Me: don't breathe. Rudy: take two steps in, dart out again.

Our ballet was Rudy's show, and week by week he hesitated less before finally slinking all the way in, lured by his food bowl sitting full and fresh inside. On cue, I released the rope and dropped the gate, then crouched far away from his bowl. We'd been doing the dance for five weeks now, and finally he'd begun to nibble his food with me in the room.

Rudy had been brought in as a orphaned kit and given a home at the center because he could not be released into the wild by law of California Fish and Game since red foxes are not native to the state. He'd lived six years at the center, spending his days lying in the grass or rooting in the sand under one of the small pink-flowering trees in his enclosure, playing hide-and-seek with the children and parents filing past his fence. Actually, he played hide-and-hide, for like a good fox he took up a spot then sat motionless, blending into the background. At night he was brought inside to keep him safe from poachers, both animal and human. Once he knew the staffperson bringing him in, he slipped into his night pen readily, but until you had passed his friend test, he tried to keep his options open.

As I waited for Rudy to calm down and begin eating, I glanced around at the gray concrete and fluorescent light. It looked like a prison. How could a fox, who loved porous, fragrant earth and potholed rock, possibly enjoy coming in at night to squared-off concrete walls and a flat concrete floor? When I later was sent with pail and scrub brush to clean out Rudy's digs, I suddenly appreciated the smooth concrete. Still, I couldn't imagine Rudy liked it. Foxes are most active at dawn and dusk, and if I'd been a fox, I'd have wanted to stay outside at twilight too.

Plus, there was the smell. Each time I passed Rudy's pen it took me by surprise—a clear whiff of skunk but not as strong. Like skunks, male foxes mark their territory with a musky secretion from their anal glands, and though the pen was hosed down regularly, every week Rudy's enclosure announced "skunk."

Still, I was safe with Rudy. Or thought I was. Hand-raised by humans, Rudy was the darling of every staffperson at the center and had never shown a hostile whisker to anyone. He was wary, like a good fox, but not unfriendly. Staffers could visit Rudy anytime we liked—unlike with the resident coyote across the way, whose pen we were forbidden to enter except to retrieve an empty food bowl. The coyote seemed mellow enough—reach your fingers through the fence to scratch his neck, and instantly his head turned toward the sky, his golden eyes glazing over in bliss. But one morning a few years earlier, when he was being walked by his favorite caretaker to the classroom to meet schoolchildren, he had reared up, snarled, and attacked the young man, gashing him badly enough to send him to the hospital.

Because Rudy seemed safe, I wanted to get closer. We were making good progress. Last week as I squatted on the floor, he had padded softly toward me, sniffing. I stayed motionless, afraid I would spook him if I so much as breathed while his pointed black nose probed the air just inches from my body.

Tonight I squatted again, this time closer to his bowl. Soon Rudy was padding toward me again and sniffing. He made a slow circle around me, one moment almost touching, the next retreating. My knee on the ground was beginning to ache. How long could I stay propped in this position? Suddenly he stepped directly in front of me. Then, without warning, he lifted one paw, put it on my lower knee, and, raising himself vertically, stared directly into my face.

I gasped. I couldn't help it. I'd never been that close to the needle-sharp teeth of a fox. Captivity, I knew, makes people, and especially wild animals, crazy. Was Rudy, like the coyote, planning to attack?

After studying my face up close for a moment, he opened his mouth, stretched out his tongue, and with it reached between my parted lips. Startled, I made an effort to keep still. After licking the inside of my mouth for a few seconds, he withdrew his tongue, let himself down to the floor, and began gobbling his food.

I'd just been French-kissed by a fox.

I was elated, I wanted to spit fast and hard—all in the same moment.

A little dazed, grinning and working grit out with my tongue, I slipped out the door, leaving Rudy to his dinner.

After that night Rudy's kisses became de rigueur. I'd bring his dinner, and once inside for the night, he'd sniff his bowl then, before grabbing the first bite, approach me where I crouched on the floor, put his front paws on my knee, stare in my face for a second, then probe my mouth with his tongue.

After volunteer duties rotated and I no longer fed him, anytime I stopped in to say hello, he greeted me in the same way. I never did get used to that gritty tongue—he'd been rooting around in the earth and in who knows what—and neither did I get used being French-kissed by a creature who smelled more than a little of skunk. But neither did I catch any dread diseases.

Only much later did I find out what Rudy was actually doing. Visiting a friend one day, I pick up a book on wildlife tracking and turn immediately to the chapter on foxes. There I read that newly weaned fox kits remain in their den while the parents go out and hunt. When one adult returns with food, the kits eagerly crowd around, trying to lick the parent's mouth, probing inside it with their tongues to see what Mom or Dad has brought for dinner and stimulating the adult to regurgitate the food.

Rudy wasn't kissing me after all; he was checking the menu. He must have wondered why I never coughed up the dog chow, dead mice, boiled eggs, and chopped fruit that appeared in his bowl.

He was also greeting me as a fox greets a trusted member of the family. Years later I still feel honored.

⁓

HAVING KNOWN RUDY, I was all the more prepared to love another fox when a tiny red kit was brought in to the hospital, mewing. He'd been found, an apparent orphan, by a well-meaning person who thought she was bringing him to the center to be cared for and raised to adulthood.

What she didn't know was the California law I mentioned earlier that prohibits releasing an animal in the state if it is not native. Red foxes were introduced to the West Coast in the late 1800s to boost the fur trade. So when the red fox kit, a potential competitor with the native gray foxes, was brought in, the staffperson on duty was required to call the Department of Fish and Game. Within minutes a van would arrive.

That day the staffperson was a young woman who loved the animals and worked for a pittance just to be near them—which is why I was taken aback when she looked at me hard and ordered, "Leave the fox alone." I stared at her, unbelieving. Was she weary of comforting volunteers who got attached to animals only to watch them die? Or was she, like I, uneasy with the law—with what was about to happen?

The kit was crying. I had to comfort him, even if my heart broke. I stepped past her and scooped him into my arms.

He was so soft!—puppy soft, but with the wiry fur of foxes. He lay in my arms, his cries slowly quieting to stillness. He slept. How could he know what was to come?

The van arrived; a man from Fish and Game got out. He strode into the building and entered the next room. The young staffer looked at me with a warning in her eyes, and I handed her the baby. She took him into the next room and closed the door. She was, I knew, placing him in the blue plastic tub, the one with two tubes leading to a tank. The switch would be turned, the gas would flow.

My arms were empty.

Was there no other solution? What was this ease in snuffing out a life?

$\rightsquigarrow$

WILDLIFE CONSERVATION TAKES place at the level of species, not individuals. Although it is illegal to kill or capture an individual native bird or animal (except for sanctioned hunting), it is not individuals that conservation laws are designed to protect. It is species. The value is thought to reside in the whole, not in the individual animal or plant. It makes sense: only a population, not a lone animal here and there, can play its role in the drama of eating and being eaten. "Occupying a niche" is the conservation biology term for how a species, like red fox or box turtle, does its part to maintain the give-and-take of an ecosystem, eating its specialized diet of rabbit or leaf, its habits rippling outward through the whole system. Lacking enough members of a species, a niche is not filled, tipping the whole out of balance.

The plants and animals have built community on a grand scale, growing up together and shaping one another over millennia of evolution. They feed on each other, they adapt to the climate together. What one lacks, another provides. When one plant suffers too much predation from a certain family of insects, another may develop pheromones to call in a different family of insects to feed off the first. Plants and birds coordinate their schedules, blossoms opening on the one day of the year that hummingbirds will stop by on their annual migrations. The plants and animals know one another; they move in a wild and intricate dance that they, over millennia, choreographed together. The dance is rugged and delicate at once, fine-tuned over eons, life devouring and supporting other life in just the right measures to keep the music alive. Individuals are born and die, but the community moves forward through time.

The view of nature as a living, pulsing community is an inspiring one, a vision of holism that in a single moment both crystallizes the isolation that most people in this society feel and shows us how to address it—by developing ties with others, both human and more than human, and relearning ways of cooperating with the web of relations that enfolds us. It's a vision I share—up to a point.

I'm just not quite comfortable with a rosy blush it wears, as if the biotic community were *only* stable or harmonious, as if life on Earth were not also unpredictable, often threatening, its rumbling-belly earthquakes destroying homes and lives, its constantly changing weather (even before humans began messing with it) demanding continual creativity, or adaptation, on the part of all life forms if they're to have even a hope of surviving.

And sometimes the correspondingly rosy visions of human community translate into simple nostalgia for old-style villages where neighbors helped each other harvest, or for the barn raisings still practiced among Amish today. When talk of community begins, and I hear a certain tone of longing creep into the conversation, I sometimes feel compelled to speak up. "I come from one of those traditional communities," I say. "Let me tell you a little about mine."

∼

IT IS THE 1690s in the valley around Bern, Switzerland. Some industrious farmers are making trouble for the government—not for what they do but for what they won't do: they won't help defend the valley. The authorities consider them traitors, for anyone who won't serve in the militia cannot possibly be a good citizen. After all, the feudal bargain has been kept since Roman times: military service to your landlord in return for the right to grow your food on his property. And if these days it is the town or canton that orders you to war, still, refusing that order amounts to being a squatter.

These tenant farmers are strange in other ways too. They keep to themselves, rarely mingling or worshipping with their neighbors in church. Instead, they meet with one another in their own homes. They refuse to swear oaths, claiming that a person's word is good enough. They call themselves Mennonites, after one of their organizers, Menno Simons, during the Reformation a hundred fifty years earlier in Holland. Their landlords are glad for the neat farmyards they keep and for the ingenuity of the ditches they dig to bring the river to their rented fields, but everyone is baffled by their stubbornness. They simply will not take up arms. Twenty years before, the governor expelled their preachers and threw dozens of people into jail. Now he is threatening worse. He plans to get rid of them for good—ship them all down the Rhine, no questions asked, give them fourteen days to pack up and get out. Families begin to pull up stakes, saying tearful good-byes to other families then sailing down the river or slipping across the hilly border into Alsace (now France), where they have friends and relatives.

Into this unsettled time comes a man named Jakob Ammann. He is young and zealous. Already in his twenties his devoutness secured him a role as an elder, one of the group of men who sets the policy for all the Mennonite congregations in the canton. Now he is dissatisfied with how the rest of the elders are treating the case of a woman caught in a falsehood. She should be banned, says Ammann. The ban, or excommunication, has been practiced by Mennonites since their beginnings, but it's always been the final, extreme, step in a process intended to turn the heart of erring members and reconcile them with the community. Ammann thinks her punishment should start with the ban, and he is taking it further. She should be shunned as well—no eating or drinking with her, and her husband should no longer sleep with her.

The rest of the elders are not so sure. Can't this Ammann leave well enough alone, especially now when the community is being persecuted? Why should the church too harass its members? Frankly, they don't like how he single-handedly excommunicates those who don't agree with him. The ban is supposed to be a communal decision of the elders, not the act of one lone preacher. They also don't like that Ammann has been calling for stricter rules—preaching against the new fashion of buttons on clothing and forbidding men to trim their beards.

The Swiss elders demur; Ammann is going too far. They refuse to follow his lead.

But Ammann persists, traveling throughout the canton, preaching to the pockets of Mennonites in neighboring Alsace. And whether he is that charismatic, or whether people are so gripped by fear—for Alsace too is harassing them—that they pull inward and set their boundaries ever tighter, Ammann wins the loyalty of many. Nearly all the Mennonites in Alsace join his cause, reverting to old-fashioned hooks-and-eyes on their clothing and practicing both the ban and shunning. They are known from that time on as Amish-Mennonites.

A hundred and fifty years later, threatened by yet another war and its backlash against local pacifists, many of their descendants will get on a boat and sail for America. Some will settle in the newly purchased Northwest Territory of Ohio, and a few will become my ancestors. Though they eventually drop the name Amish and again call themselves simply Mennonites, they will continue, into the twentieth century, to excommunicate and shun those of their number who commit misdeeds. Every time the government raises an army they will suffer harshness—taunts about cowardice from their neighbors or repercussions from the law—and in every war they will turn this harshness toward their own transgressing members, setting tighter rules and policing their boundary with greater vigilance. Their community discipline will rival that of any army, and they will be known for their impeccable morals—as stiff, upright, and plain as the vertical planks siding their barns.

⁓

To outsiders, Mennonites have always looked very different from the rest of society. Even though they long ago gave up wearing old-fashioned clothing, they continue to be confused with Amish—for good reason, I learned when I found that my branch of the Mennonite tree in fact grew on Ammann's side of the fence. (The Amish-Mennonites of my heritage are but a fraction of Mennonites worldwide. Today Mennonites, like Quakers, actively work for peace and justice and to relieve hunger and suffering in local communities.)

To my insider's view, though, the Amish-Mennonites of one or two hundred years ago look a lot more similar to modern folk than anyone imagines. I think, for instance, of their male elders, handing down decisions as efficiently as any board of a college or corporation. True, the elders wielded an authority that today is unheard of, setting policy on everything from clothing to farm implements. Certainly no freedom-loving individualist Americans would tolerate being told what to wear or what tools to use!

Or would they? I think of the colors we will buy next summer, chosen by fashion designers deciding what will be "in," or tech companies rendering our computers or phones obsolete almost before we switch them on. We are under great pressures to conform—a different pressure, to be sure, than the bishop reminding a woman at the church door that her sleeves are too short, but a conformity that may be more insidious just because it is so invisible. On plenty of matters we moderns seem willing to lay down our own wishes to pick up instead the desires laid out for us, as a wardrobe, by others.

Author Tobias Wolff claims we have an "appetite for the 'mindless contentment' of self-surrender," a nearly insatiable hunger for which we have arranged "a singularly rich offering of oppressions to satisfy." He counts on the smorgasbord everything from fascist philosophies to mass-market advertising to a psychology of determinism that tempts us to discount our own free will. It's an astonishing menu of ways to give up personal power. From his list you might think we had no imagination at all for dreaming up satisfying ways of organizing ourselves or for shaping humane forms of authority—modes of power that enable, not hobble, individuals.

I mean, could a people truly committed to individual well-being possibly have dreamed up the assembly line?

~

IT IS THE summer after my first year of college. I stand on a rubber mat on a concrete floor. In front of me is a long, winding table covered by a conveyor belt. Overhead, fluorescent tubes glare with a bluish light, turning the skin of the person across from me a pale gray. As particleboard shelves inked with wood grain approach me on the belt, I reach for a rag with my ungloved hand, hit the spring-loaded lid of the lacquer thinner, wipe excess ink from the shelves, and slip them into a cardboard carton. This I do from 7:00 in the morning until 3:00 in the afternoon, five days a week, my life regulated by buzzers loud as fire alarms to mark the beginning and end of the day as well as the thirty minutes set aside for lunch.

At my end of the floor, cinderblock walls stretch as far as I can see in all directions. In the middle of each long wall sit several windows, bright squares of warm light and green leaves announcing that summer is taking place on the other side of the wall. In between moving cartons, I stare out the window, calling to mind the memory of how grass feels between bare toes. My coworkers dream of Friday night, when they will collect their paychecks then blow a good share of them at one of the several bars in town.

The standing and the boredom and the toxic fumes take their toll on me as well. By the end of the summer my blood pressure is dangerously low, and I have trouble standing up for the whole day. Once I get home in the afternoon, all I can do until bedtime, my body aching, is to lie on my back on the living room floor with my legs propped up on the seat of a chair.

~

A DECADE AFTER meeting the red foxes, I am attending a yoga class—breathing, moving, sensing, breathing. In ten years I have learned much about biotic communities. Along urban creeks I have planted hundreds of native plants and pulled hundreds more invasives. In my neighborhood I have become something of a conservation expert, teaching others about why native plants make better choices in their yards than pampas grass, French broom, or ivy.

The yoga class is winding down, and we settle on our backs into shavasana. My mind, as usual, darts to and fro. I breathe, and breathe again. Minutes pass while silence slowly descends throughout the room. Finally there is quiet inside me as well.

Suddenly, unbidden, an image arises. It is the red fox kit from a decade before, nestled again between my arm and heart, his meek mewing for his mother slowly quieting. I feel again the appreciation I offered him, as if I could make up for what would happen next.

He was born in the wrong place; he encountered humans; he died. A sadness for him grips me, a deep-reaching powerlessness I haven't felt in a decade.

I know that gray foxes are just as fetching; this is not a matter of cuteness. I don't want the grays squeezed out, just as I don't want the delicately furred native blackberry vine of California creeksides smothered by the fierce-thorned Himalayan blackberry or the native bunchgrasses on the hillsides choked by the spiny star thistle (poisonous to horses). These are species that don't play well with others. How can they? They grew up—evolved—in another part of the world, with a different set of flora and fauna shaping their development. Introduced to North America, they lack the give-and-take of a community that knows them, plant and animal neighbors to keep them in balance. In North America the newcomers simply take over.

But what about the overwhelming irony of this law? Of conservation by eliminating nonnatives? The rule that decreed the red fox kit's end was written by governments that trace their heritage to Europe. From that same land came thousands of colonists for reasons that had little to do with promoting diversity. Most were not interested in playing well with those already here. They confined and killed the natives; like star thistle or red foxes, they simply took over, making the land their own. My ancestors, arriving in northwest Ohio just as the Indians were herded onto trains bound west for reservations, helped to decimate the deciduous wetland that had covered the area since the last ice age.

I attend a writing workshop in Arizona, where a Navajo woman, tears of fury squeezing out the corners of her eyes, chokes to all the white people in the room, "Sometimes I wish you'd never come here!"

⌒

KILLING, AS A solution, comes easy to this culture. We seem to reach for it first. Is there a problem we haven't tried to solve by wiping it off the face of the Earth? Pesticides, herbicides, war, antibiotics—other options considered only if eradication doesn't work, which it usually doesn't, at least not in the long run.

A hundred years ago the federal government, freshly honed from the Indian Wars—not wars but genocide—took aim at its next target: predator animals. Hunters around the country were clamoring for more deer, and ranchers wanted their cows and sheep to roam freely across the range. So the Bureau of Biological Survey, part of the Department of Agriculture and a forerunner to the Fish and Wildlife Service (and its California counterpart, the Department of Fish and Game), sprang into action. It set out to rid the country of predators—wolves, coyotes, mountain lions, bobcats, bald eagles, and bears.

Most of the large animals had already been taken care of by burgeoning human populations, and in many areas only coyotes remained. Between 1915 and 1947, the years of all-out campaign, almost two million coyotes were killed, most with bait drops of strychnine-laced horsemeat. To makers of conservation policy, this was effective "management" of the country's "resources."

Killing as a solution makes sense if you look at nature as a machine made up of parts that lack their own purposes and so can be rearranged to suit our own. (Language of "managing resources" always reveals the contours of a machine.) And in this case the machine enjoyed a spectacular but very short success. In one "managed" area, the Kaibab National Forest on the north rim of the Grand Canyon, deer numbers shot up as intended after only a few years of predator elimination. But as deer multiplied, they also gobbled up all edible grasses and plants, devastating the landscape, then began to sicken with malnutrition and disease. Their die-offs were even more spectacular than their flourishing—tens of thousands dead in two winters alone, between 1924 and 1926.

By the 1930s the idea of eliminating all predators fell out of favor; the national parks would become small, enclosed sanctuaries for predators as well as other animals. But by that time the Biological Survey had cultivated what amounted to a specially trained army of hunters and poisoners, and groups of these "conservationists," angry at the new policy, would camp outside the borders of the parks and sneak in with guns and poison to bag their kill. Predators would still be exterminated systematically on other public lands, with an ever-increasing budget, which by 1971 amounted to $8 million. The large-scale poisoning itself would not end until the 1970s, replaced by sharpshooting from helicopters, which still continues today along with federal trapping, hunting, and poisoning programs.

And where does control by killing stop? If you "manage" the predators, don't you have to "manage" also the animals they feed on? In an interrelated biocommunity, how can you stop with one species?

Thomas King, of Cherokee descent, writes, "If you believed in such a world," where some have value and others do not, "there would be no end to the killing."

‿

ALDO LEOPOLD SUPPORTED the killing. Leopold was a forester in the 1920s who loved to hunt. Even after those around him were beginning to voice their doubts about the blanket killing policy, Leopold continued pushing for it. He appreciated its efficiency. As late as 1933, in his textbook on game management, he wrote that effective conservation requires "a deliberate and purposeful manipulation of the environment," meaning, in practice, picking off predators (though by that point he too opposed blanket extermination). The forests, rocks, and rivers of the western country could be cultivated in the style of eastern agriculture, to produce "crops" of sheep wool and cow meat for ranchers, and equally plentiful "crops" of deer for hunters.

It would take a personal meeting for Leopold to come around to a different point of view.

One afternoon Leopold, on forest patrol in rimrock country, spotted a group of wolves down below at the river. As usual, he opened fire then scrambled down the rocks to claim his prize. He arrived to find the lead wolf, a mature female with a litter of grown pups, still alive but unable to move. Fascinated, he moved closer to her, holding his gun ahead of him as a shield. With her last ounce of resistance she snapped, clamping her jaws onto the butt of his rifle. He backed away and watched as she slowly died.

Years later he would write that he saw "a fierce green fire dying in her eyes."

I realized then, and have known ever since, that there was something new to me in those eyes—something known only to her and to the mountain. I was young then, and full of trigger-itch; I thought that because fewer wolves meant more deer, that no wolves would mean hunters' paradise. But after seeing the green fire die, I sensed that neither the wolf nor the mountain agreed with such a view.

It's as if Leopold's point of view suddenly packed up and moved outside himself, coming to rest in the mountain or in nature itself. In truth, the process took decades; he met the wolf in his early twenties, and his "Thinking like a Mountain" was written more than thirty years later. But at the end of his life he gave that one wolf the credit. She started the process that over years transformed him into someone who taught us to see ourselves as part of something larger, part of the land-community.

The "land ethic" is what Leopold finally called it, an ethic of acting in ways that benefit the soil, trees, animals, and insects who made this world. The land ethic has one simple guideline: "A thing is right when it tends to preserve the integrity, stability, and beauty of the biotic community. It is wrong when it tends otherwise." Following Leopold, people found it easier to adopt a more-than-human perspective, as if we slowly realized that preserving nature for its own sake will—of course—benefit us, who depend on it. We are one "humming community," in Leopold's words; we are one body.

The dying wolf was not the only one who prodded Leopold toward this view. In the 1920s he read the newly translated book of a Russian mystic and animist named P. D. Ouspensky, a book that was taking American intellectuals by storm. "There can be nothing dead or mechanical in nature," wrote Ouspensky. "If in general life and feeling exist, they must exist in all." After reading Ouspensky, Leopold wondered if the Earth was made for human use or if perhaps we got that idea all backward.

> *Possibly, in our intuitive perceptions, which may be truer than our science and less impeded by words than our philosophies, we realize the indivisibility of the earth—its soil, mountains, rivers, forests, climate, plants, and animals, and respect it collectively not only as a useful servant but as a living being. . . .*

Today hardly anyone remembers that Leopold once considered Earth a servant; he is known instead for moving a whole society away from trying to conquer nature and toward becoming a "plain member and citizen" of the land-community. And he brought about a seismic shift in national forest management. The old aim of squeezing products out of forests—what Leopold called growing "trees like cabbages" for their cellulose—was out. In its place came public lands managed as natural areas, as examples of what Leopold called "land health" and these days is called ecosystem health. More than four decades after his death, in 1992, the U.S. Forest Service officially turned its bureaucratic ship in this direction.

And here's where Aldo Leopold's story connects with that of a tiny red fox kit brought in to a wildlife rehab center in northern California a few years later. For Leopold's vision of "one humming community," his unwavering commitment to valuing all the parts of nature's body, had blossomed into a movement to preserve biodiversity. So when the governments, first of California in 1991 then of the nation the following year, turned toward preserving ecosystems, they instituted practices to preserve native species. Nonnatives would have to go. They might not be poisoned or killed as systematically as before, but once captured, they also would not be released back into the wild. The little red fox's fate was sealed.

⌒

I FIND MYSELF astonished by the paradoxes. Nonnative humans giving the gas to equally nonnative animals—an irony so glaring that a friend who is as white and Euro-derived as I, after hearing the story of the fox kit, wonders aloud, "So, what's next—putting ourselves in that blue plastic tub?"

Not to mention our stubborn habit of managing animals as if they were parts of a machine to be switched on and off at our whim.

But the paradox that interests me most, at least here, is that a faith in Earth-as-community could emerge at all in a culture that has such a limited—and limiting—view of community. It shows up even in Leopold, even in his land ethic. You have to read closely, but there it is in his opening paragraphs: "An ethic, ecologically, is a limitation on freedom of action in the struggle for existence."

He explains: individuals when they follow their instincts are in an "original free-for-all competition." It's the old notion of competition as the driving force of evolution, and there's a lot more to be said about it, but let's follow Leopold for now. Though we are compelled by instincts to compete, he says, we can learn to cooperate by following ethical guidelines. Instincts may drive us apart, but ethics can bring us together.

I find his words odd—and jarring as well. If instincts are what separate us, then by nature we are antisocial. In other words, community is not natural. You have to override instincts to make it happen.

Jarring it may be, but the roots of the idea go all the way back in Western history to Plato himself. The passions, those unruly instincts, he said, are like horses that must be held in check by the driver of the chariot, the mind. Cultures that trace their history to Plato tend to be convinced at their core that civilization happens only when people rein in their natural desires and inclinations or competitive drives so they can get along.

There's a lot to be said for this view. We all know that we can be selfish. We don't act in every moment as if the interests of others were as important as our own. The problem comes when selfish or antisocial is *all* we think we are.

It's a pessimistic view of our own nature—as so out of whack that we need outside help to bring us into line. The community becomes a referee, like a parent who must intervene to keep the kids from fighting over dinner.

Is it any wonder we have trouble finding community? Trusting community? We have set it up as a check on individuals, whom we imagine as no better than squabbling children.

～

ABOUT THE TIME the red fox kit appears in and then too rapidly disappears from my life, I attend a workshop led by a West African couple, Malidoma and Sobonfu Somé. This is the Malidoma whose book I will edit a few years later and whose experience of speaking with a tree will come to seem unremarkable to me. But all that lies in the future; I attend because at the moment I simply need comfort. I am still grieving, unable to shake the persistent sense of loss. Though several years have passed since my divorce and the deaths of my parents, and now and then an hour or a day gleams brighter than before, most of the time my life feels thieved by sorrow. I am withdrawn, stuck.

The workshop that day seems meant for me: its theme is grieving. The hundred of us, divided into smaller groups, build altars around the room to the elements of nature. In the corner devoted to fire, my group and I fashion a structure blazing with votive candles and aluminum foil and swatches of red and orange fabrics, a swirl of color and light. Then we sing simple African chants and dance to the beat of drumming that lingers through the afternoon as each of us collects our own griefs and presents them at the altars of nature.

But the thing that stands out most from that day is what Malidoma and Sobonfu said at the start when they tried to explain why grieving is so important—why people in the Western world experience so much grief without recognizing it and why we need to take time to feel and release it.

"In our village," they said, "when a woman becomes pregnant, everyone looks around and wonders, Who is this person coming to join us?" Villagers assume that the person is bringing gifts from the ancestors that the village needs. They speculate excitedly to one another on what those gifts might be.

As the pregnancy continues, a ritual is performed to answer their questions. The elders meet with the pregnant woman and, through her, speak with the fetus. "Why are you coming here right now?" they ask. The village learns what hopes and dreams this person-in-the-making brings to the world—above all, what gifts the child, once here, intends to deliver to the community. Will she have the fire of the ancestors burning brightly in her so that she can be a beacon to the village when it cannot see the best way forward? Will he be peaceful, like water, to smooth the rocky places between people? Will she be skilled in shaping gifts of the Earth into baskets or pots? Will he be fluent in the language of drums?

It then becomes the job of the community to remember that person's gifts after she arrives. "We all forget our purpose," Malidoma said, for we go through the rigors of being born then living as infants and toddlers. As we grow we need to discover our gifts anew.

"The community exists," said Sobonfu and Malidoma, "to help individuals remember their purpose."

I sat bolt upright, astonished. This was a view of community I had never heard before. My experience was the zero-sum game of American and Mennonite societies alike, a tug-of-war between individual and community over the ultimate prize, freedom.

I could hardly fathom—and I believe most Americans cannot imagine— a community that pays special attention to each individual, nourishing each one with the quality of attention that helps each bring forth her deepest gifts, as if every community member were engaged in doing what too often is done only by parents or the best teachers. And all because the adults know that every child will make a difference—not in some abstract way, but *for them*, for the world they will make together as the child grows. As if each adult were thinking, while gazing upon each child, "This person brings me—us—a priceless gift."

It's a different view of nature—of our natural state as bringers of gifts, each of us a contributor, not a barrier, to community.

Sobonfu explained what it looks like in practice. In the village, when a child enters the room, the body of every adult turns toward her to welcome her, to give her their full attention. Children need this complete attention, she said, if they are to learn to respect and trust their own gifts.

Is there a practice more different from the American norm, where children who make noise in public usually receive irritated looks from nearby adults? As if, in the twenty-first century, children still should be seen and not heard.

What would it feel like, as a child, to be nourished with full-hearted attention? To know that every adult around you has a personal interest in the gifts you bring into the world? To have the community feel responsible for helping you sort out your dreams and find the best creative path you possibly can?

$\rightsquigarrow$

WE CAN PRIZE the gifts within only if what lies within us is trustworthy.

But in the Western world we suffer from centuries of believing something else—that what lies within us, and within all of nature, is broken, flawed. It needs help. Plato's chariot with its out-of-control horses fed this idea, as did, of course, the Genesis tale of Adam and Eve. But in the Roman Empire of the fourth century one man pushed it further.

Augustine was just twenty-six, at the start of a promising academic career, when in the year 380 Christianity was made the official religion of the empire. A half-dozen years later he was on top of the world—professor of rhetoric to the emperor himself in Milan. Here he could teach and argue and shape the thinking of an empire. But Augustine was stricken, deep in a crisis of faith. Manichaeism, a popular spiritual movement with its vision of a cosmic struggle between good and evil, could no longer hold him, and neither could mystical Neoplatonism. He needed something more immediate, more personal, something to heal the sickness and guilt he felt so keenly within. The new official religion provided the relief he had long sought. He converted to Christianity then, in one fell swoop, gave up his prestigious job, sold his possessions, distributed the money to the poor, abandoned both fiancée and longtime mistress, and moved back home to what is now Algeria to be ordained as a priest. From his post in North Africa he would become the most influential bishop in the empire.

Over the course of his career Augustine devoted a good share of his powerful intellect to combating one particular nemesis, a man named Pelagius, said by some to have come from Ireland. Pelagius at first seems an unlikely target. Born the same year as Augustine, Pelagius had been educated in Rome and now, like Augustine, was an ascetic and preacher. But Pelagius had chosen a riskier path than priest. In Rome he saw what the first flush of political power was doing to Christianity, and he didn't like it. The new Christian aristocracy was ignoring the needs of poor people and living wanton lives of luxury. Pelagius set out to reform the Roman congregation.

So far he and Augustine were on the same page. Both were celibate, following strict morals and preaching purity. What they disagreed on was how goodness comes about. Pelagius thought we have the power to choose it. He refused to blame nature for human failings. True, we are attracted to wrongdoing, he said, but only because of the force of habit, and habits can be changed. Pelagius, in other words, was an early self-help guru; he believed in working on yourself.

Augustine was alarmed. He felt powerless in the face of his own failings; that's why he had converted to Christianity in the first place. No, he said, we do *not* have a free choice. We are wounded at the core, and in spite of our best efforts we lose touch with our more admirable selves. We need help.

Augustine had the weight of four hundred years of Christianity behind him, though Christians were hardly alone in believing this. I think for instance of the religion of Isis, Queen of Heaven, which had been thriving during those same centuries. "Isis saves!" sang her devotees joyfully in temples across the Mediterranean; they, like most people in the Roman Empire, looked forward to ultimately escaping from this clearly not-divine world. More pessimistic still was the Confucian sage Xunzi, who in China hundreds of years earlier taught that any goodness found in us is learned and not innate. Xunzi believed too in authoritarian government, as those who emphasize inborn evil often do; the vice within must be held in check by an outside force.

In the face of Augustine's opposition, Pelagius went on preaching. "It is up to us to make better choices," he insisted. "It's not nature's fault." Call him prescient: he saw clearly that to say we are powerless against evil implicates all of nature in a serious way.

Augustine too was prescient, but he took away a far different lesson. If people could be so responsible on their own, what need was there for the Church? Now that Christianity had risen to power, there could be no room for Pelagius and his hopeful ideas about self-improvement. Augustine came down hard on Pelagius, and the Church councils backed him. Pelagius was condemned as a heretic and excommunicated. Original sin became the starting point for the law of the land: all humans tend toward evil, a trait we inherit just by being born.

It was a theological glove fitted to an imperial hand. The Church from now on would be the sole conduit of grace for the millions newly within its fold. Can it be an accident that it felt compelled, in the heady days of its new power, to assert human sinfulness?

Certainly it is no accident that, even if Pelagius was not Irish, his teachings took root throughout Celtic regions of Britain and Gaul, where animism still held sway. There the centuries-long belief in the goodness of the Earth, including the goodness born in humans, died out only slowly.

~

AFTER AUGUSTINE AND Leopold, after assembly lines and authoritarianism, what sense can we make of the red fox kit and the law that sent him into oblivion? I suggest that when we think about biodiversity and ecosystems—when we imagine nature as a community—what goes into our thinking is not just details of fox habits or hummingbird migrations or nectared flowers or even all of them put together. In the shadows of our minds lurks this history, this idea about what community needs to be—a restraining force—because nature, especially the nature inside us, is broken and needs fixing.

After Augustine it would take a thousand years, but eventually this vision of a deficient Earth would shape itself into a vision of Earth as a passive thing, a machine with parts that can be switched on and off at our whim. Plants and animals would no longer speak; they would toil away in their own little corner of nature's factory with instincts loud as factory buzzers telling them when to eat, when to sleep, and when to return to their jobs. We, the operators of the machine, would take great pains to keep them under control, for hadn't we decided long ago that those of us who crawl on Earth are but fractious, sinful creatures in need of help?

Which is why the most hopeful worldviews today—the ones that can guide us toward friendlier relations with the wider community of nature—are found in cultures that do not share this history. Malidoma and Sobonfu's people, the Dagara of Burkina Faso, share with many Africans a vision for goodness flowing from individuals to their communities. The Ghanaian elder and Canadian sociology professor George Sefa Dei says, "To many Africans, the dichotomy is not between the *individual* and *community*." Instead of individuals as competitive units separated from one another, as in the Western world, in traditional African thought individuals are cooperating, "enriched by community." Nourishment streams always in two directions, from individual to group and back again.

Where plants and animals can speak, it is common for human individuals also to be valued highly. After all, if every rock or bush or rabbit may be a conversation partner, then living well means listening to unique voices, including human ones. I have heard American Indians from various traditions talk about how much room there is among their people for difference of opinion—more, they say, than in mainstream white culture—because the Creator may say different things to different people, and all messages deserve respect.

An anthropologist named Thomas Buckley recently told a story about living among the Yurok people in northern California, along the lower Klamath River. After dinner one evening, while sitting beside the fire, an older man picked up a stick and asked Buckley what it was. "A piece of firewood," Buckley answered. The old man looked away, disappointed.

Buckley tried again. "It's wood, a piece of a tree." This was better.

"And what's a tree?" the old man prodded.

The conversation stopped. But that night, as they were turning in, the Yurok man offered more.

"When you can see each leaf as a separate thing, you can see the tree. When you can see the tree, you can see the spirit of the tree. When you can see the spirit of the tree, you can talk to it and maybe begin to learn something."

Each leaf as a separate thing. Exquisite particularity. The individual as the path to communing with the whole.

⌐

IN THE EYES of the law, the fox kit was out of place, a weed, expendable. To me, his individual life was precious, his death arbitrary.

I desperately wanted a different solution. But in the fifteen years since, I haven't been able to think of a good one. Release him? Where? In California, to threaten the gray foxes? Or ship him—even if there were money for such things—to his historical territory in Canada or Colorado? And how well would he do there? Once cuddled by humans, a small fox can find it hard to live again as a fox—too trusting of humans, too ready to override his internal warning system.

Or suppose there had been an open slot at a local rehab center, and he could have lived Rudy's life. Plenty to eat and lots of love, but at what price? To never explore open spaces, never be allowed to cavort in happy circles under starlight or probe a fox mouth with his tongue—is that what I wanted for him?

And so I come to the end of the fox kit's story, no closer to a solution than I was at the start. As I am puzzling over the pieces of our combined stories, a friend asks, "What would the fox kit want you to say?"

The question startles me. I have no answer.

But within a few days I have a dream. In the dream I am close to the ground, padding on four little black feet next to another young fox who is leading me somewhere—across damp rocks toward someone at the far end of the murky cave. There she is, lying down, her full teats ready. We dive into her side, pushing and slurping, hot and urgent with our need. The sour-sweet of her belly is home. It is safety. It is life.

And then I understand. Deprived of *her*, how would I live? How *could* I live? So confusing it would be to learn to drink from a bottle! To be held by dangerous-smelling hands! Between a life so alien and a death as easy as breathing in, there's a lot to be said for the latter. I might even choose it myself. Especially if it was preceded by cuddling.

But at the time this end was chosen for him by others, I could not feel such equanimity. My bag of griefs, already heavy, sagged even more with the weight of the baby fox.

Something in the law seemed wrongheaded. I hated its deadly efficiency. I felt depressed by its consequences.

5

Stories We Live By

The truth about stories is that that's all we are.

—THOMAS KING, *THE TRUTH ABOUT STORIES*

T HE SUMMER OF 1995 I was thirty-eight and at a standstill—a dissertation that never ended and depression humming like a motor that someone forgot to turn off. Though despair was less raw than before, I still spent most days in isolated silence, slipping away when I could to hike on deserted beaches and tending birds once a week at the wildlife rehab center.

I yearned to get away, maybe take a vacation, but I hardly knew where to begin. How do you plan a holiday if you've never taken one? Besides, how could I possibly afford a trip? I couldn't even afford to stay home. My small inheritance had evaporated during long afternoons spent napping to recover from CFIDS, and each semester's tuition payment was bumping up the credit card balances. Now that I was well, it was time to get on with my life, but dissertation and finances seemed determined to hold me back.

Maybe it was time, like George on *Seinfeld*, to do the opposite of everything I was used to doing. What could it hurt? I would leave town for a few days and simply trust the money to work out.

One thing I knew: I didn't want to travel solo. This trip would have to be a vacation from loneliness as well. I researched guided tours and finally landed on the perfect match: a women's travel company out of Seattle running a simple three-day camping trip to Mount Rainier in late July. The price fit my graduate student budget. I reserved a place and booked a flight.

The camping trip provided everything I'd hoped for. Being cooked for was heaven enough—organic fruit, gourmet sandwiches, sizzling breakfasts. But then there was the mountain itself. We trooped up trails that wound through dense stands of conifers and across roaring streams and finally, on the second day, arrived at the snow line, where we scooped up wet globs and tossed them at each other, laughing. Our lunch counter was sun-warmed rocks, the view extending for miles while at our feet alpine wildflowers glowed white and yellow and indigo. Through binoculars I spotted, far above, trekkers headed across the slope, a line of ellipsis dots on an ice-white page. On our way back down the mountain, we came around a bend to find ourselves at the top of a waterfall that disappeared down a chasm of rainbow-speckled light.

The trip provided everything—everything, that is, except bald eagles.

It was when Anya invited me to her house on Lopez Island that my trek out of loneliness truly began.

⌒

I SHOULDN'T BE surprised that the meeting that invited me into a new kind of life took place on an island. Islands are self-contained, or so they appear, separated from each other by often-forbidding waters. They are exemplars of isolation. And yet we have that saying too that none of us is an island, all separateness an illusion. As nature philosopher Kathleen Dean Moore says, "not even an island is an island," for every spot of land rides atop the deep fabric of Earth and becomes a sign, "a beautiful, rock-solid, bird-spattered sign, of the wholeness of being, the intricate interdependencies that link people and places."

Yet interdependence gets short shrift in our guiding stories—the stories that tell us, at bottom, how the world *is*. In religion they're called creation stories, renamed by secular society as myths. But secular society too has its origin stories announcing how and what to think about the world.

～

"IN THE BEGINNING was the atom."

It was a story told by Democritus and his teacher, Leucippus, in Greece around 400 BCE. It's not the creation story we usually tell, but it might well be.

"The atom was separated from all other atoms by a great void."

In truth, Democritus and his teacher probably didn't tell stories. Myths and poetry were the old-fashioned way, used by Homer and the bards to celebrate a living cosmos. Democritus and his generation were done with all that; they wanted precepts and principles, not stories about how the world worked.

And so our isolationist tale took hold—first in ancient Greece then again at the dawn of the scientific revolution in northern Europe two thousand years later. It is a story of particle-islands swimming in a formless, empty sea, a creation story that has taught its Western listeners to value separations and distinctions and to view all of nature through the eyes of isolated individuals.

In sixteenth- and seventeenth-century England, where atomism took hold for the second time, the mechanistic philosophy that it made possible allowed emerging science to cast off the power of the feudal Church. But just as in ancient Greece, where atomism replaced the older story of a living Gaia, so in early modern Europe atomism displaced a belief in the mysterious workings of Earth. Nature died at the dawn of the scientific revolution, not because there was better evidence to support the new story but because the old story had been monopolized by a Church that hoarded nature's powers, and especially land, to augment its own authority. For intellectual—and economic—freedom to flourish, religion had to go.

Or at least a certain kind of religion.

In one of the more spectacular acts of tossing the baby with the bathwater, English philosophers in 1600s began to purge not just religious authority but also religious feeling from their thinking. The process would continue in the Enlightenment of the following century; by our own time, scientific inquiry would take place within a rationality defined as absent of reverence and awe. Intellectual sophistication would come to be equated with believing that the Earth is impersonal and dead, its atoms propelled by mechanistic forces that can be completely described by human reason.

It is a sort of scientific Puritanism, and in fact many of the seventeenth-century men who set it in motion were Puritans railing against "popish superstition." To them, superstition meant whatever excited the imagination, especially during worship. They had already torn out the popish stained glass from their churches and evicted all paintings and images from the sanctuaries. No glowing candles or warmly colored light would distract the faithful from a sober focus on the Word. It was literal truth the Puritans wanted—plain, unadorned truth. The truth disclosed in words.

Already used to plain thoughts in church, the scientifically minded applied a Puritan aesthetic to their intellectual life as well, seeking a view of nature unsullied by florid superstition. It was a stark, even ruthless vision they pursued—nature as seen in the cold, hard light of day. Their scripture would be mathematics, a form of writing that for the scientific faithful would yield insights as powerful as those of sacred writ, and in the centuries to come, their trust in its texts would rival any Protestant's in the Bible.

༳

"A SPOKEN STORY is larger than one unheard, unsaid," writes Chickasaw poet Linda Hogan. "In nearly all creation accounts, words or songs are how the world was created, the animals sung into existence."

If words can bring worlds into being, what kind of world have we created from the story of atoms separated by a void?

And what sense can we make of the fact that we have uttered words intended to erase stories, not create them? The scientific revolution was used to strip the animals and plants of subjectivity, to turn them into objects ready for human use. For, as Francis Bacon put it in 1620, "Knowledge and human Power are synonymous," and the purpose of knowledge is to further "the empire of Man over Things."

How can we understand this urge to sing the animals *out of* existence, to take away their stories?

༳

A TIMID MAN, Thomas Hobbes abhorred conflict and shrank from violence. He had plenty of chances to cower, for he lived in England in the chaotic 1600s, a century filled with religious wars, social upheavals from the emerging system of markets, and political feuds with other states. Hobbes would eventually run for safety at the first whiff of chaos, fleeing the Civil War of the 1640s when the Puritan-led Parliament rebelled against the popish King Charles I. This was the same Charles who a few years before had set up a draconian court that cut off the ears of Puritans and branded their bodies to punish their heretical views.

Hobbes, more horrified by social unrest than by any tyranny of the king, supported the monarch. He escaped to France to live with the exiled royal family and kept himself busy by tutoring the young Prince Charles in the mathematics he had recently learned from Galileo. At the war's end he published his new book, *Leviathan*, a defense of strong government.

Society needs strong rule, argued Hobbes, because the normal state of affairs is so terrifying. Without absolute authority, chaos rules. Human nature cannot be trusted. Or rather, it can be trusted only to lust for power: "a perpetuall and restlesse desire of Power after power that ceaseth onely at death." The human compulsions to rip all possessions from neighbors and seize power are so strong, thought Hobbes, that society, left to itself, runs amok in what he called a "warre of every man against every man."

Then the sentence for which Hobbes is famous: without external restraints, life is "solitary, poore, nasty, brutish, and short." This, he claimed, is the "state of nature," common to all animals as well as to "the savage people in many places in America" who "have no government at all."

Hobbes was wrong, of course—wrong about there being no law among American Indians as well as about animals endlessly fighting. A little close observation would have told him as much. But empirical observations would take center stage in science only later; for now first principles ruled, and Hobbes's first principle was isolated, warring individuals.

Hobbes can be forgiven for thinking that all was violence and thievery. After all, a land grab had been going on already for a hundred years before he was born: manorial landowners seizing fields formerly farmed by their serfs or tenants and enclosing the land in pastures for the sheep whose wool fetched a hefty price in the suddenly opening foreign market. Poor farmers were displaced, whole villages sometimes destroyed, and England was suffering the effects of the upheaval, including an uptick in theft and violence. Decades before Hobbes's birth, Sir Thomas More had criticized the practice of enclosure, blaming it for much of the theft. But during Hobbes's lifetime, as foreign markets continued to reward exports, the practice only increased. Beginning with landowners and ending with peasants, the world had turned voracious, with people determined to seize anything belonging to their neighbors.

There is, Hobbes thought, a natural check on all the violence. Because everyone is equal, and equally violent, eventually individuals come to an uneasy truce. Self-interest balances out if all pursue it equally. The better way to check the violence, though, is through government restraints. Let everyone consent to an absolute authority whose job it is to keep people from destroying each other, then all can pursue their own interests freely without fear. This is the social contract.

It was a dismal take on human nature, and many of Hobbes's contemporaries were appalled by it. Not until after his death did Hobbes's story take hold, preparing the ground for Adam Smith's *Wealth of Nations* a century later. Hobbes's warring individuals would become in Smith's version the happy individuals who pursue their own self-interest and so bring about social harmony. But for Hobbes and all his successors, society is made up of atoms, separate individuals competing for the goods of life. The individual precedes society in Hobbes's story; disconnection lies at the core of his universe.

As capitalism took hold, Hobbes's influence would grow, leading his jaundiced view of human nature to haunt Western thinking down to our time. And here's the thing: it made sense not just because it harmonized with the mechanistic views of the new philosophers and not just because it explained the fevered acquisitiveness of the newly prospering merchants but also because it rang true with ancient traditions of Christianity—with the creation story of Genesis, which was working alongside the Greek story of isolated atoms to bring modern Western society into being.

Hobbes, who despised the Puritans for bringing about anarchy, had ironically absorbed the most basic tenet of their Calvinist faith—that human nature is fallen. His "state of nature" matched in every respect what Calvinists of his time would have said about Adam and Eve in Genesis— not Adam and Eve in the garden, but Adam and Eve after they sinned: they lived a cursed existence and passed their depravity on to all their descendants. Human beings were powerless to improve their lot.

The Genesis story of the fall had taken center stage in Protestant churches for at least a hundred years before Hobbes because Luther and Calvin had made it a set piece in their Reformation. God's grace, they proclaimed, was free to all—no need for pope and priests as the conduits. But to be able to seize upon grace, people first had to be convinced of their sinfulness. And so Luther and Calvin had repopularized the Augustinian notion of original sin. In what became the great irony of the Reformation, Protestants could separate from the late medieval church only by reviving Augustine, that pillar of the early medieval church. Original sin would become the starting point for the Reformation—then the starting point more than a century later for the social contract theory of Hobbes.

Hobbes may not have been religious, but he too was convinced of fallenness; he just called it the "state of nature" instead.

Those in our time who hold a similarly pessimistic view of human nature (usually political and religious conservatives) tend to favor, as Hobbes did, strong government control, and in recent years Hobbes has enjoyed something of a renaissance. The *Internet Encyclopedia of Philosophy* praises him as "an acute and wise commentator of political affairs" and admires him "for his hard-headedness about the realities of human conduct." During the administration of George W. Bush, Vice President Cheney, a defender of strong executive power, stated that Hobbes was his favorite political philosopher.

Hobbes has had a lasting influence on economics as well, with his "hard-headed" view defining the starting point of Western capitalism. Society as a whole has come over to Hobbes's way of thinking, content to believe that acquisitiveness lies at the core of nature. His pessimistic view lives on, even among those who might be astonished to learn of its religious roots.

୰

THE IDEA THAT organisms compete with each other to survive has become so synonymous with evolutionary theory that if you dare to question it, people assume you are rejecting evolution itself. And though Darwin did think competition was the axis on which evolution turned, his story of nature did not end there.

According to Darwin, evolution proceeds by natural selection. But exactly what this means has been hotly debated ever since *On the Origin of the Species* came out in 1859. In general terms it's obvious: natural selection means the environment rewarding those traits that give animals or plants greater ability to survive and reproduce. Natural selection is how traits become more or less common in a population depending on whether those traits help or hurt survival on that particular spot on Earth.

(By contrast, consider genetic drift, the idea that traits become more or less numerous through accident alone. In this model traits have nothing to do with survival ability; they're not selected by the environment for survival. Other models of evolution emphasize cooperation—for instance, symbiosis, in which two unrelated organisms combine to form new kinds of life. Evolutionary biologists now know that symbiosis among both bacteria and viruses has shaped human evolution.)

But the exact "how" of natural selection has never been fully explained, not by Darwin or by his successors. Darwin himself pondered it for a very long time before committing to one theory. And when he did commit, it was because he was grabbed by the writings of a quiet country parson turned economist, Thomas Malthus.

୰

MALTHUS WAS IN his early thirties and newly ordained when in 1798 he decided to counter what he considered an alarming trend: a growing optimistic belief that society could be made perfect. Only twenty years earlier Adam Smith had published his *Wealth of Nations,* which would become the bible of Western capitalism. Smith began with Hobbes's bottom line: the idea that humanity is fallen, that people tend to act in self-serving ways. But Smith accomplished an extraordinary feat. With his hopeful theory of markets, Smith managed to transform this flaw of selfishness—normally something that drove people apart—into the glue of social cohesion. If each man pursued his own self-interest, all people would benefit. What before was morally suspect—acting to gratify one's own needs without considering the needs of others—became acceptable behavior. Even greed, thanks to the "invisible hand" of the market, might be transformed into the engine of public good. (Never mind that Smith never explained how the magic happened—nor has anyone else—and never even asserted it outright though he has been used for two hundred years to argue for it.)

Smith's optimism was infectious, buoyed by the profits that merchants and gentry were enjoying thanks to their plunder in foreign lands, their new industries, and their takeover of formerly common rural lands. Monarchy had fallen in France, and new technologies promised endless improvements. No doubt poverty and disease could be eradicated as well.

But Malthus felt there was more to the story. Malthus hastily printed a dissent, calling attention to the misery that was increasing alongside the wealth. Displaced from rural lands, people had flocked to towns and cities hoping to find industrial jobs, but low wages and unemployment meant that masses now lived in poverty. Malthus could not pretend that the new order would result in infinite good, and he thought there was a logical reason for it.

Using mathematical models, Malthus argued that population growth is destined to outpace food supply. Human population, he said, increases geometrically while agricultural resources (food) can increase only arithmetically. Nature will never keep pace with human demand. Poverty is inevitable. What therefore ensues is a "struggle for existence," a fight to the death over the available resources.

Like Hobbes, Malthus started by assuming ruthless competition. But instead of criticizing unequal methods of allocating resources, he chose to blame the poor. Hunger could be solved if poor people limited their family size. For, after all, the parson Malthus argued, the deeper causes of social problems lie within people, not outside them. In fact, a lack of food might not be all bad, he suggested, for human beings are at their core sinful and sluggish, and only food shortages can motivate them to work for a living.

As for Augustine and the Reformers and Hobbes, so also for Malthus, the fault lay in nature itself, in the nature outside human beings as well as inside them. These were "deeper-seated causes of evil," Malthus said, found in "the laws of nature and the passions of mankind."

⁓

WHEN DARWIN READ Malthus, a light went on. The last piece of the evolutionary puzzle fell into place, namely, the *how*: evolution took place through natural selection defined as a struggle for existence. Nature did on a grand scale what humans did in breeding and gardening: it selected for the "favourable variations" and destroyed the "unfavourable ones." Competition, as Malthus had understood it, was the logical mechanism.

Darwin's world-changing insight occurred some months after he returned from a five-year voyage to South America, where he had grappled with harsher landscapes than those he knew in verdant England. On the Galápagos Islands, west of Ecuador, he had discovered nature sculpted from ash, its volcanic structures big yawning maws of black rock, its creatures not the woolly cud chewers of home but overgrown lizards and strange finches feeding on stunted, spiny shrubs. He confessed to being rather horrified by the place. It reminded him of Staffordshire, "where the great iron-foundries are most numerous." Like the towns of the industrial revolution, these were islands of isolation filled with arid, blackened earth, where life was harsh, uncompromising, and born of struggle. Herman Melville, visiting the Galápagos only a few years after Darwin, lamented, "In no world but a fallen one could such lands exist."

And yet on that blasted landscape Darwin was converted from a geologist to a biologist, for the varieties of life on those islands were so astonishing he had to take account of them. What mechanism could give rise to such diversity? For more than a year after returning home, Darwin studied feverishly, trying to make sense of it.

Then his epiphany:

> *In October 1838, that is, fifteen months after I had begun my systematic enquiry, I happened to read for amusement "Malthus on Population," and being well prepared to appreciate the struggle for existence which everywhere goes on from long-continued observation of the habits of animals and plants, it at once struck me that under these circumstances favourable variations would tend to be preserved, and unfavourable ones to be destroyed. The result of this would be the formation of new species. Here then I had at last got a theory by which to work. . . .*

And so the theory of ruthless competition, born at a time when humans were casting about for ways to make sense of the ruthlessness they were practicing among themselves, passed into biology to help explain animal and plant life as well. Darwin accomplished a monumental task: by proposing common descent between humans and other creatures, he reidentified us with the rest of the natural world. He healed, at least in theory, the split that the Western world had opened between itself and the rest of nature. To this day both science and religion tend to shrink from the full implications of that insight.

But by locating his mechanism for evolution in the competitive individual, Darwin made Hobbes and Malthus come to seem natural as well. Nature itself looked remarkably like selfish human beings scrabbling their way past others in a struggle for existence that was "red in tooth and claw," as Tennyson so famously put it.

Even Darwin sometimes shrank from the sight. In the same month that he achieved his breakthrough, he scribbled in his scientific notebook, "It is difficult to believe in the dreadful but quiet war of organic beings going on [in] the peaceful woods & smiling fields." Yet believe it he had to, even here in his own serene backyard, for what else could explain those turbulent lands he had seen on his voyages?

After Darwin, the dog-eat-dog view of nature would be endorsed by the emerging field of biology, and deviation from it would eventually be scorned as beneath intellectual dignity. Perhaps biologists were reeling from the implications of being so closely related to "beasts," for in the late nineteenth century they used natural selection to retreat into a dogma that resembled Descartes more than Darwin—animals as mechanisms driven only by mindless instinct. The dogma would harden into Pavlov's idea of conditioned reflexes, where animals would be studied, as Pavlov said in 1927, only "as physiological facts," without any "fantastic speculations" about their inner states.

Yet the dogma rested on a certain dishonesty: some feelings would continue to be attributed to animals, but only those that fit the jaundiced view of Hobbes and his successors. Antagonism and competition between animals would be assumed, while affection and kindness would for the most part go unnoticed. Recently, primatologist Frans de Waal reports that a colleague gave a professional presentation in which she dared to frame baboon activities in terms of friendship and received a cool and skeptical response from the audience. "Can animals really have friends?"

he says, "was the question of colleagues who without blinking accepted that animals have rivals."

What de Waal terms our "Calvinist sociobiology" culminated a few decades ago in the theory of "selfish genes," which enthroned selfish need as the prime mover of all creation. The picture of a callous and unfeeling nature was complete. It would be a secular version of the doctrine of a fallen world, its roots stretching back fifteen hundred years to Augustine and his vision of nature pursuing its sinful course—a conquest of the Western mind to a degree that Augustine could never in his wildest dreams have imagined.

For such a triumph, would he have wept with joy?

I RIDE DOWN a country road on Lopez Island. A bald eagle flies in from the edge of the island straight toward my car, floats in low circles overhead, then wings its way back to the spot from which it appeared—all this after I had sat quietly, calling eagles with pictures and with love.

Coincidence? The mechanistic theory tells us it is. The other story, from Genesis as filtered through Augustine then the Protestants, chimes in that such connections belong to a time "before," when the world was more perfect than it is now.

Two creation stories with roots in the ancient Mediterranean—one proclaiming islands of matter bumping randomly, the other lamenting a tragic world that is not how it was intended to be. The stories dance together down the centuries, flirting with each other, as stories do, climbing in and out of bed, eventually moving in together for good in early modern Europe.

Not that they were the only two stories in the house. I think of that other Greek idea, from Plato, that a bright and perfect world exists in a place only the mind can go. During the Roman Empire, this ancient Greek mistrust of matter mingled with the Christian version of the fall, sealing Augustine's belief that the physical world is sinful. But Augustine had been influenced too by yet another story, a Persian one about the cosmic struggle between a spiritual world of light and a material world of darkness, a story that to this day gets revived whenever big troubles are brewing.

But it is the story of isolated atoms and the story of the fall, two nearly opposite tales, that found each other and shacked up—as it turns out, happily—in northern Europe at the beginning of the modern era. That's the thing about opposites: they often get along famously, probably because in most ways they're so very much alike. And in this case, though they disagreed loudly about whether the world is moving toward some purpose—the mechanistic one barking, "Of course not!" and the religious one patiently repeating, "Yes, all will make sense in the end"—at a deeper level they had nothing to argue about. Each knew beyond the shadow of a doubt that intelligence (for the mechanist) or God (for the religious one) was completely divorced from the natural world. They were alike as two peas in a pod in their belief that the human mind is the only place where awareness, intellect, or spirit may be found on Earth.

Joined by their conviction that humans stand above and separate from nature, the mechanistic and religious stories went about eroding any remaining connections between humans and Earth. To this day they teach skepticism about the natural world over trust in nature's processes. Through their influence, dissatisfaction with nature flows throughout Western civilization, as deep as its blood, as abiding as its bones. Convinced to the marrow that something is deeply wrong with nature, Western people have set out to fix it—both the outer nature of impenetrable forests and untillable wetlands, and the inner nature of the sinful human heart. Discontent with "what is," the Western world tries to remake it into something better.

Together the stories became the ancestors of most modern Western institutions, including its scientific method, its commerce and government.

If forests or mountains or rivers get sacrificed in the quest for endless profit, the stories work together to convince people that it can't be helped because, as Malthus said, the deepest evils lie within. Ask the average person today how to solve the problem of greed, and she or he will likely say, "You can't. It's just human nature." Most people have forgotten that it was only a few centuries ago that laws were made to reward private accumulation— and that those laws were written and supported by people convinced that nature, and especially human nature, is selfish and acquisitive at its core.

Most have forgotten as well that the choice to reward only one half of our nature is just that—a choice, or rather a series of choices that continue to be made to this day. Believing that "baser" impulses rule nature, Western cultures arrange their members into adversaries poised for endless competition. Setting people up as opponents creates the need for strong external power, and so in Western cultures the larger community is given the job not of loving and nurturing its members—for love on the large scale is impossible if nature is selfish at its core—but instead of setting limits to the harm they may do to one another.

In biology, the story of mindless struggle takes center stage, while evidence of altruism or cooperation among creatures until recently was ignored. Today the idea that animals act out of blind and selfish instinct still seems to much of the public, and to many scientists, more "value free" or scientifically accurate than the idea that animals may be acting out of love, connection, or morality. Yet it took hundreds of years and some clear theological trends to arrive at this version of reality, one that would be considered ludicrous by most traditional or indigenous societies.

Why the premium on "value free"? they might ask. And what kind of people would consider such a vision of reality, lacking love or empathy, to be true?

ↄ

THOUGH I DIDN'T realize it at the time, mistrust of nature was on trial that day on Lopez Island. Beaten down by grief, unnerved by isolation, I was testing the stories handed down to me. Is the world bleak, as portrayed? Fallen, or friendly?

Or is nature perhaps, as David Abram put it, "not 'nice' but far more beautiful"?

ↄ

AN EAGLE'S TALONS can crush with a strength ten or even twenty times that of an adult human hand—up to four hundred pounds per square inch. Bald eagles are magnificent. They embody grandeur. Gliding low over a lake, they grasp a fish they may have spotted from a mile away, lifting it cleanly and carrying it off to eat.

And yet that day on Lopez Island, a bald eagle spotted me from beyond the horizon and flew out of his or her way to carry me, not toward that oblivion of light, but nevertheless across a threshold, away from one life and toward another.

ↄ

"SO FAR AS I am aware," writes anthropologist Marshall Sahlins, "we are the only society on earth that thinks of itself as having risen from savagery, identified with a ruthless nature. Everyone else believes they are descended from gods."

Of course, other cultures too recognize human flaws—our constant jockeying for status; our lust for power and control, for lovers we can't have, for anything just out of reach. But most cultures also recognize another part of our nature. They resist the urge to think that greed and selfishness must always win. When love and affection and altruism are seen as rooted just as deeply in nature—in *our* nature—life can be arranged around sweeter values.

"Two matching birds sit in the exact same tree," says the several-thousand-year-old Mundaka Upanishad. One bird is the self, with a small *s*. It thinks it is isolated from others, and so it is miserable, "remaining stuck in feelings of grief and powerlessness." The other, the Self with a capital *S*, sits in the same tree serenely watching, for it sees that the source of love and joy lives within every creature, even in "a clump of grass." It is not caught up with striving and sadness.

We have a choice which bird to identify with. We can go, of course, with the isolated one. It's just as real as the other, and they do look so much alike! But if you follow this one, says the text, be forewarned: your life will remain hemmed in by winning and losing, and you will feel that you are never quite making it.

Instead, "Identify with the Self, not the self, and grief will fly."

It's a radical solution, just the opposite of the one my culture chose several hundred years ago, of scooping up all the goodies for yourself because it's the "law of nature." Starting in a different place, the Mundaka Upanishad recommends equally different action: "Those who see the source in all creatures no longer strive to outdo others. They rest in the peace at the center."

Is this a viable path—for more than just spiritual dreamers? I can just about hear the scoffing: Don't you know the real world follows different rules?

Yes, some "real worlds" do, especially the one that Europeans dreamed up, to our planetary peril. When love abandons the world, or when people choose a story that says it has, ruthless competition rushes in to fill the void.

Which is why I feel a jolt of disorientation, a welcome jolt, every time I hear of a society that organizes its real-world life according to a different story.

ↄ

LIFE IN THE Andes is about nurturing and letting oneself be nurtured—a startling premise to those of us whose story begins in competition. Though I have not visited the Andes, I am listening to what native Andeans are saying, and they are telling a story that has sustained the Quechua- and Aymara-speaking peoples and dozens of other indigenous Andean cultures for centuries.

I know that cultures are like individuals, marching down through time with their own peculiar blind spots, their weak points and contradictions. No doubt the regional Andean culture has its warts as well, but for sheer longevity and peaceable cooperation with nature, its success is impressive. Given the tough conditions in which Andean people live—steep mountains, an austere climate at ten thousand feet—they could have chosen a story of nature as harsh and uncompromising, an adversary who must be wrestled for food. Instead, they chose a different tale.

In the Andean story, humans as well as llamas, potatoes, maize, beans, ancestors, rains, and mountains all keep life running smoothly by loving and allowing themselves to be loved. The very Earth depends on it, they say. Without affection and respect, crops cannot grow and life will not regenerate. People will become sick and fields infested with pests. Sweet relationships are the foundation of life itself.

Nurturing others is pleasurable, they say. It is not an obligation. You don't give love so you can get taken care of in a tit-for-tat trade. You give affection because it is simply more enjoyable than the alternative. It deepens intimacy. It enhances relationships. It shows you that you are *not* self-sufficient, and this is a source of happiness. In this complex web of interrelationships that is Mother Earth, or Pachamama, everyone needs affection from others.

Andean people nurture the plots of ground, or *chacras*, of maize and vegetables, and the soil reciprocates with crops. But the llamas too are a *chacra*, the "*chacra* that walks," as the peasants say. When llamas are well tended, they nurture humans with their fur. Humans too are a field, a *chacra* nurtured by deities, who walk alongside rather than directing life from some heavenly podium because they too are limited; deities also need love and affection. An atmosphere of "profound equivalency" guides all relationships—between humans and nature, between humans and deities, between deities and nature. "In the Andes we are all people and we are all related," says one writer, and life "emerges from conversations between similar and equivalent beings." Each watches for cues about what will keep the others happy, and each gives the affection and respect the others need. "With them we keep company," says another; "with them we converse and reciprocate."

Within the living Pachamama, who encompasses stars and wind and sky, not merely ground, nothing is inert. Pachamama lives because she is regenerated by the affection flowing among all. When relationships are sweet in the household, Pachamama is healthy and life will be renewed. When disharmony or deception prevails, Pachamama is injured and food is less likely to grow.

As in any household, conflicts do arise, and then affected parties come together to listen and speak. Deities and nature beings might speak through divination rituals, while human beings speak their hearts. Once relationships are reconciled, life can flow freely again. Because everyone in nature is speaking, one has to listen closely, moment to moment. Those who do not speak verbally speak always through signs and actions. Is the wind arising stronger than expected? What is it saying? Are there pests in the maize fields? What type of nurturing does the maize lack that it has become unbalanced? "Conversation is thus an attitude," writes an Andean native, "a mode of being in unison with life, a knowing how to listen and knowing how to say things at the appropriate moment."

As I read the Andeans describing their way of life, I wonder how smoothly it all works in practice. Do they get the listening-and-speaking right every time? Probably not, if life in my own corner of the world is any clue. People often talk past one another, and the more our communications depend on words, the more likely misunderstandings are to arise. But if the Andean practice of trying to live in unison with life looks idealistic to Westerners, how much of our perspective arises from a deep-seated pessimism about what is possible, either among humans or with nature? How firmly are we convinced that social harmony lies beyond reach just because "that's the way things are"?

*

THE SENSE OF activity, of interactivity, must be profound in Andean daily life. I would love to witness it. Imagine the feeling of aliveness—that everything you can see (and everything too that you cannot see) is moving, responding, swirling, affecting its environment moment by moment, always shifting, doing, reacting, creating. And all this moving and making and fashioning and forming can happen smoothly when each part gets the love and attention it needs.

As I read about Andean culture, I find myself thinking back decades to those years when my own life force was precarious and I dwelled in illness. The remedy lay not in adversarial medicines like antibiotics and antivirals or in toughening up my emotional skin but in their opposites—in nurturing my body with love and attention, in following every impulse toward kindness, in leaving relationships that depleted and seeking out those where affection flowed freely.

Perhaps when life is on the edge, as in an illness—or a changing climate—only love and nurturing define the line between survival and perishing.

On Lopez Island, while I walked through the valley of the shadow of death, I needed nurturing from anyone who could provide it. Of course I was providing it too during that time, tending the injured and orphaned wildlife at the rehab center, nurturing them with food and attention. But on Lopez Island I nurtured the eagles in a different way, with conversation. I courted them with binoculars and bicycle and, yes, with begging too.

And because the flow of life depends on sweet relations, the eagles responded in kind. One even nurtured me with an in-person visit.

"Here we are!" the bald eagle said. "We heard you."

*

PERHAPS, TO PARAPHRASE David Abram, the world is not nice but far more intricate. As robust yet delicate as friendship. As thorny as politics or partnership. As messy and blissful as marriage. Family dynamics are tricky, after all, and living together peacefully requires some effort. Respect. Conversation. Negotiating. Being willing to balance one's own needs with those of others. A playing field of "profound equivalency," as the Andeans say.

Which means that the idea of using nature only to satisfy human needs, the foundation of the Western economic system and now the hallmark of the global system, is not just bad environmental policy but also bad politics. If the world is made up of relationships—nothing but relationships— then our whole political economy requires an overhaul. For the Earth to renew, life will need to become sweeter, with more give-and-take, more affection, more nurturing.

The need for the overhaul is urgent, and easy to see, when the life that is being devastated is one's own. At ten thousand feet in the Andes, even small climate changes have catastrophic results. Andean people are well aware of this, so they are bringing to global climate talks a "Universal Declaration on the Rights of Mother Earth." It begins with the idea that Pachamama is a living being, "the source of life, nourishment and learning." Earth and her inhabitants, they say, have the right to clean air and water and to freedom from toxic contamination. Andeans are writing into their national constitutions similar rights for nature, such as the rights of all beings to exist and regenerate and be free from tampering by mega-structures that upset nature's balances. Ecuador instituted such rights in 2008 and Bolivia in 2011. It is a bold strategy for standing up to foreign corporations that extract local resources; it is a bold attempt to preserve the health of the local Earth.

It is also an intriguing case of instituting animist philosophy at the national level. To those who think that public life can run only on the mechanistic story, Andean culture and political life tell a different—and centuries-old—tale.

~

THOUGH DARWIN BELIEVED natural selection was ruthless, his story of nature, as I said, did not end there. He remained awed by all that humans and other animals shared, including feelings. In this way evolutionary theory challenges the mechanistic story as well as the religious one, for no longer can animals be viewed as simple machines running on instinct. (But because Darwin was one of the few to absorb this insight, after his death the emerging field of biology snapped back toward the Cartesian view as if on a rubber band.)

When he was a young man, Darwin speculated in his scientific notebooks (1837) that humans and other creatures "may be all netted together" as "fellow brethren." But after decades of observing animals and thinking about evolution, he became much bolder. In *The Descent of Man* (1871) he asserted outright that humans and many others share emotions and feelings.

> *Man and the higher animals, especially the Primates, have some few instincts in common. All have the same senses, intuitions, and sensations—similar passions, affections, and emotions, even the*

more complex ones, such as jealousy, suspicion, emulation, grati-
tude, and magnanimity; they practise deceit and are revengeful;
they are sometimes susceptible to ridicule, and even have a sense
of humour; they feel wonder and curiosity; they possess the same
faculties of imitation, attention, deliberation, choice, memory,
imagination, the association of ideas, and reason, though in very
different degrees.

The champion of natural selection never forgot that nature had selected not just for competition but also for compassion. Instincts include a
sense of fairness and generosity, an ability to make choices.

How far down the evolutionary line does such intelligence extend? In
his final book Darwin set out to answer the question by studying earthworms. He staked out observation posts at the burrows of worms and took
copious notes about their daily habits. By counting, measuring, and ana-
lyzing, he discovered that earthworms make some sophisticated choices.
For instance, in handling the leaves with which they plug their burrows,
they tug on the leaves 90 percent of the time from the tip end rather than
the stem, easing their work because the tapered tip slides more easily in
dirt. They plaster their tunnels with the earthworm equivalent of cement
and build little baskets at the doors of their tunnels to rest in and shield
themselves from the chill of the winter earth. In short, they show agency
and choice and therefore "some degree of intelligence." They act not out
of pure instinct but out of a certain degree of mind, "in a manner as would
a man under similar conditions."

The article that outlines Darwin's work with worms bears the pro-
vocative title "The Inner Life of Earthworms." The writing, except for
the title, is careful, cautious, as Darwin himself was. As I read it, I am at
first delighted—what a clear refuting of our own reductionistic age! Then
almost immediately, the dismay—what an impoverished awareness, that

we need such painstaking arguments to "prove" intelligence in other creatures, something that might be obvious if we were not so committed to a radically different story!

⟿

IF EARTHWORMS MAKE their world, and by extension ours too, what sense can we make of all this making? Perhaps reality is like an interwoven mesh of crisscrossing trails, every being an actor, every "thing" creating something new in the world. The continuously creating world is a theme that crops up repeatedly in the writings of indigenous people. From the sacred hoop of the Lakota to the "continuous web of creation" of the Maori, the cosmos in many cultures is a circle of continuous making, everybody a creator, everybody enmeshed with all in crafting an ever-evolving world.

It's a vision of reality that Western physicists began to glimpse a hundred years ago, as relativity theory and quantum mechanics challenged the distanced, objective view of reality that had reigned in science since the seventeenth century. In contemporary physics, "you can't reduce the subject to an object," says physicist and professor Arthur Zajonc. If the observer is always implicated in the results of any observation, as quantum physics shows, then reality is "irreducibly subjective. It's irreducibly participatory." Physicist Karen Barad even coins a term for it: *intra-acting*. Instead of a world where separate elements "interact" with each other— one thing bumping into another thing, the foundation of the mechanistic view—some interpretations of quantum theory suggest a more radical picture: a world where "things" do not exist at all until *after* they interact. There are no "things" prior to relationship; each is brought into being through "intra-acting," through relationship.

Instead of Descartes's "I think, therefore I am," we might say, "Interacting with others, I am."

In this view, reality—nature, all of us—is a dynamic, swirling cacophony where trees, microbes, and humans too are brought into being through the weave of relationship. Moment to moment, we—sea horses, zinnias, and clouds—create one another. Together, all beings fashion the world.

~

MOST OF US at some point experience a deep connection with an animal. Sitting with a pet dog or cat, a place inside us may open, a place of sweetness or delight. But if the creature is a wild one, its life not contained and protected by humans, we may feel something rarer—the quality of gift, the sense that this other, of its free will, chooses to be present with us, to be, if only for a moment, a companion.

In that moment our perception may shift. The self we thought we knew—maker of worlds, the species acting on other species but only rarely acted upon—drops away, outdated. For we see, in a moment, that this other also has its own will, makes its own choices, perhaps loves also, or grieves.

After that moment of seeing, life is no longer the same. A new self emerges, a self we gain as we grant a new status to the other. This new self sees how much of life it shares with the other—experiences of observing and choosing, curiosity and risk, delight and pain. It may see too that the choices the other has made—about where to build a nest or where to sink roots or how to raise a family without a mate—have made an impact on the world, have helped to create the world, our world.

When the bald eagle approached and flew overhead, I was forced to see not just a friend but a cocreator. In responding to my need, the eagle brought something new to life. Some*one* new, that is. I was born into a new way of being.

Like all infants, I would grow only gradually into this new skin. It would be a years-long process of paying attention in new ways, of growing into trust for the impulses of wind and water, of practicing connection instead of isolation, of training my eyes to see and my heart to feel, of learning to see the Self in all creatures, even in "a clump of grass."

But for the journey I soon received a guide. Only two months after I met the eagle, someone new entered my life.

6

Sapphire

When love awakens in your life, in the night of your heart, it is like the dawn breaking within you. Where before there was anonymity, now there is intimacy.

—JOHN O'DONOHUE, *ANAM CARA*

"LOOK WHAT I did in a moment of weakness!" Esther called as I got out of the car. "Wouldn't you like a dog?"

My landlady was standing in our yard one Monday afternoon a couple of months after I returned from Lopez Island. I had spent the day hiking on Mount Tamalpais just north of the city, keeping my promise to myself to walk among trees or surf one day a week to see if it would help my still-ragged insides. Now Esther was waiting at the fence along with her brown Doberman, Tango. But next to Tango was a dog less than half his size, black mottled with white, lean and rippling with muscle. The new dog was studying me quietly out of one blue eye and one brown as I opened the gate to join them.

"I couldn't stand it when I heard about her," Esther said. The day before, Esther had attended a picnic with a woman who worked at a pound thirty miles away. "She said they had this dog there, and all the staff just loved her. They moved her around from kennel to kennel—you know, keep her out from under the supervisor's nose. She'd been there six whole weeks." Unclaimed animals usually lasted only days. "But I guess he finally found out about her. She was scheduled to be put down today." Esther's face clouded.

First thing Monday morning Esther had climbed into her car, driven to the pound in the suburbs, and paid the forty-dollar ransom. "I think she's kind of funny looking," Esther added, almost blushing. She hated to say an unkind thing about any animal. "If you want her . . . "

For months Esther had been trying to convince me to get a dog; Tango needed company. I had plenty of reasons to say no. I was a graduate student. Not enough time. Definitely no money. What if there was a medical emergency? "I don't think I can," I said, though I bent to greet the dog. Those eyes were striking. She stood still—friendly, curious—while I petted her oddly mottled fur. "But I *will* help look for a home for her."

I tried. I called my birding friend Meredith, who lived at the other end of the bay with an aging dog and two cats, and told her of our boarder. Meredith said her house was full, but she promised to call others.

The next day, as long as the dog was here, I decided it wouldn't hurt to give her some exercise—take her along on my afternoon walk in the wild-land parks above our house. I flipped down the backseat of my hatchback, spread a blanket in the rear, and expected her to jump in and pace around, as dogs do. Instead she found the open passenger-side door, leaped nimbly into the front seat, and parked herself upright, gazing out the windshield. She looked so pleased with herself that I couldn't bear to dislodge her. On the way to the park she didn't hang her nose out the window, like other dogs. She just calmly watched the road in front, turning to look out the side window only for something really compelling, such as a dog dashing down the sidewalk.

At the park we headed down my favorite trail. "Heel," I told her softly, tugging briefly at the leash to show her where I wanted her to walk. It took only two or three tries, and she was there. But then she gave me one long look and deliberately walked a few steps ahead. Yet she didn't pull at the leash or lunge at people or dogs. She was calm and alert—good trail sense. So I unhooked the leash. She waited a moment until I was ready then trotted on ahead, matching my speed. We hiked a three-mile loop through the hills this way, she moving brisk and alert, reaching the next bend in the trail then pausing to look back: *Are you coming?*

Her decision had clearly been made.

That night, wavering—maybe a pet would be good for me after all—I invited the dog into my basement apartment, and she slept on a blanket on the floor near my bed. As dawn broke the next morning, I heard her stir and stand up. Then I felt her gaze bearing down on me like a weight; she was ready to get on with her day. But since I was doing my relaxation practice, floating between dream and waking, I remained still with my eyes closed, and she, after watching me for a few moments, quietly returned to her blanket, sank down, and curled up again. I loved her then—her quiet sensitivity, her being willing to adapt to my habits.

That night I invited her onto my bed, and for the next decade she slept nowhere else, our bodies breathing side by side while she draped her chin over my ankle or shoved her curled-up spine hard against my hip.

～

ONCE THE NEW dog had moved in, it was time to find her a name. I was impressed by her steady calm, and "Serena" almost won out. But there was something steely—no, that's too cold a word—something rock hard yet gleaming, like jewels, just below the surface of her. I tried to express this to Esther, and she simply said, "Sapphire." We looked at each other and nodded.

Indoors Sapphire at first was tentative. She seemed perplexed by furniture and only stood in front of the futon sofa, gazing with longing at its soft cushion. I didn't mind her lying on the sofa, but I wanted to see if she would get there by herself. After a few days of wishing and hoping in front it, she slowly lifted one paw and placed it gingerly on the seat. Catching sight of me watching, she quickly pulled back her paw and smiled an anxious, one-sided grin. I said, "Go ahead. You can sit there," but it was another day or two before she dared to climb up.

In the house she was unfailingly polite—quiet and sedate, never chewing shoes or pulling books off shelves or biting furniture or carpet. The vet said she was only about two, yet it seemed as if normal young-dog behavior was beneath her. I got the feeling she would rather have died than disturb something that wasn't hers. During the day I would sit at my computer desk, while across the room she slept on the futon, a serene, uncomplicated presence.

After she'd been with me a few days, at a certain point in the afternoon she would hop down from the futon and lie on the floor behind me, staring at my back. She had a stare that accomplished things—a gift from her herding bloodline. Her gaze would bore into me, a patient, silent command. It was time for our afternoon walk.

Outdoors she was a different dog. In the park she came alive, pouncing at lizard holes, dashing zigzag across a hill, inviting other dogs to chase her—and always outracing them—or spinning a hundred and eighty degrees in midair for the sheer joy of it. She was quick, lithe, lighthearted. I felt my heart lift just watching her.

But her free spirit outdoors was not without its drawbacks. As soon as she got to know me, she began to drop her perfect off-leash behavior. Instead of simply running a few happy-crazed circles then coming when I called, she began dashing headlong into the underbrush, twigs snapping, after some rabbit or mouse. Over my frantic calling I would hear her crackling trail recede into the thicket. And that look she gave me!—a full-on stare in the act of plunging away. Fifteen or twenty minutes might pass before the thicket would crackle again, far away then closer, and she would burst onto the trail, panting, an aroused, ungentle look in her eyes.

I was worried. What if she ran onto the road? What if she got lost? I needed to break the behavior before it became a pattern. I tried being stern with her, sitting her down quietly as soon as she returned and giving her a lecture. She merely turned her head away, looking confused. She knew I didn't like her running away, but that jewel-like surface was impermeable. To make matters worse, I soon found I was catching poison oak every time she ran off, and I had to give her a bath. She hated baths, getting so upset after the first one that she threw up her dinner, though after that she merely stood, ears drooping, through the wretched routine.

So a few weeks later I made an appointment with an animal communicator named Winterhawk, who had been recommended by someone at the wildlife rehab center. I needed her to talk to Sapphire: How could we solve this running-away problem? After meeting the bald eagle, I was willing to give this talking-with-animals thing a chance. Even with someone named Winterhawk.

I called Winterhawk at the agreed-on time. Over the phone her voice was warm and cheerful. She explained that she communicates with an animal through pictures. She sends a picture to the animal then pauses on the line to see what pictures appear in her mind in response. I waited a few moments as Winterhawk silently introduced herself to Sapphire. Then came Winterhawk's voice: "Wolf! Wolf! I am wolf!" I wondered what that had to do with anything, until I remembered I had scheduled this visit to talk about Sapphire's habit of running away.

Winterhawk then asked Sapphire about her name. She liked it, very much. Next was diet—how did she like the food I was feeding her? "Fresh meat would be much, much better." (Of course: wolf. We added fresh meat to her meals.)

But in talking about off-leash behavior, Sapphire was completely uninterested. Winterhawk spoke silently with her for a few minutes, showing how dangerous it is near roads, with human cars and all, but Sapphire seemed unimpressed. "The hunt is more important." I got the feeling she needed to run in the underbrush the way I needed chocolate before my period. So I settled on a compromise: I would let her off leash only far away from roads and only once in a while so she would not degrade the land. She'd get a hunting fix now and then, and she'd have to suffer only an occasional bath.

Then I wanted to hear more about Sapphire's past. Could she talk about what it was like being at the pound for so long? I waited a few moments while Winterhawk posed the question.

"I always knew that someone would come for me," Winterhawk reported, her voice so grave, so earnest, that I smiled. It was Sapphire's tone, all right.

The next instant tears jumped to my eyes. It was a lot more faith than I could claim. My life was a daily worry of rent, food, tuition, and grief, my days strongly tinged with sadness. I wondered at this dog's sense of trust.

How could she know someone would come for her?

*

SOME SAY I should have trained Sapphire better. More than one person informed me over the years that if Sapphire ran off against my wishes, it meant she had taken over. We needed to make sure that I, not the dog, was in charge. "You need to be dominant," they said. "Dogs need that structure."

Well, some dogs may need more structure. But Sapphire chafed under training. She easily learned commands then chose at once to ignore them. If I scolded her, she got that confused look and turned away. She was instead training me: the softer the tone and the gentler my manner, the more closely she paid attention. If I raised my voice even a notch, it distressed her; she shied from it. During training she seemed bored—so much better to be out exploring together! So I kept training to a minimum, only what was needed for safety. She learned to hike "on the right" instead of wandering into the path of people and bikes, and then her training was done.

I had to agree that Sapphire didn't need better manners. She was unfailingly polite. When meeting new people, she watched them with interest, waiting to make up her mind. If they stopped to pet her, she immediately leaned lightly against them. She was patient with young children, standing quietly while they draped themselves across her back. Only once in the decade we shared did I see her take an instant dislike to someone, a brash young guy at a barbecue who thought it fun to get in her face and bark. Sapphire fixed her steady eyes on him and bared her teeth, yapping, and I grabbed her collar and pulled her back—but silently applauded her good judgment.

With other dogs Sapphire was playful, not aggressive. One day we bumped into another woman on the trail with a medium-size black dog, and both dogs and humans hit it off. When we visited their ranch some weeks later—I'd been asked to consult on the woman's book—Sapphire and Daisy were in heaven, dashing headlong from one end of the property to the other.

I couldn't have asked for an easier dog-companion. Sapphire came with a built-in sense of decorum.

⌐

IN RECENT YEARS it has become popular to equate dog training with dominance. The "dog whisperer" of TV fame, Cesar Millan, has had a lot to do with this confusion. Dogs are wolves, he says, and wolves live in packs. In his theory pack animals need a leader, and it's up to the human to be that leader. You do this by showing your dog through body language that you are dominant. You go through doors first. You make the dog wait a moment at the dish before eating. You set the pace. This shows a dog his or her place in the pack and sets the dog's mind at ease. When you are calm and assertive, you are rewarded with a calm-submissive dog.

There's a lot to like in Millan's methods. He stresses calmness and structure. He mixes boundaries with love and affection. You can see he truly cares about the dogs and their people and wants to make life easier for them—especially for the dogs. At the end of each episode, clients rave about how he helped them live more happily with their animals. I have a soft spot in my heart for anyone who works that hard for dogs and especially for building peaceful relations with them.

But there are also problems with his theories. The most glaring is that wolf-pack theory has been thoroughly debunked by animal scientists. Wolves, like all other mammals, live in families, not packs. (Only rarely do two or three families combine into a pack, and even then it is likely an extended family, the lead females related to each other.) There is no "alpha dog" keeping others in line; instead, there is a leader pair, who are the mother and father of the rest of the group (though families sometimes absorb an unrelated member or two). The mother and father are the only breeding animals in the family, and they make sure none of the rest procreate. They monopolize the food—to promote their own breeding—but they do not make all the decisions, train the young, or walk ahead of the others. The older offspring of the family help raise the newest litter and also do their share of the hunting. In other words, Mom and Dad protect their reproductive dominance, but that's about as far as their dominance extends. When the young are one or two years old, they disperse, setting out on their own to find a mate and a territory. Most young wolves grow up to become part of a breeding pair, leaders of their own families. In other words, Millan's theory of the alpha dog is based on some seriously misguided notions about wolf culture.

Not to mention that little about wolf culture is transferable to dogs. Fourteen thousand years (at least) of sticking close to humans means that dogs are predisposed to pay attention to people in a way that wolves are not and never will be. Dogs have been shown to respond to human glances or body language with a similar degree of sensitivity as human infants. Wolves lag far behind, even wolves who have been hand-raised by people. The physiology may be close, but in social organization dogs are no longer wolves.

Yet what interests me here is not Cesar Millan or his dog theories but rather how he skyrocketed to success. His television show in its very first year, 2004, became the most popular show on its channel. He has sold millions of books, and one report says that fully half of Americans recognize his name. His "leader of the pack" theory has spread everywhere, copied by trainers and experts alike. Even a highly respected humane society I later came to know included in its adoption packet a small book teaching new dog owners how to be leaders of their pack.

Why did his message take so readily? It's obvious that Millan is personable, self-possessed, and media savvy. I'm sure that the calmness and compassion he often models added to his popularity. But I'm just as certain that those values alone would not have won him such a huge following. Plenty of dog trainers teach gentle methods, and some of them have written books and hosted TV shows. They simply don't get as far.

What Millan did was brilliant: he took what the American public already believed—that animals naturally struggle for dominance—and draped it in a new mystique, that of "dog whisperer." The title came from the 1998 film (and earlier novel) *The Horse Whisperer*, about a trainer who uses his preternatural affinity for animals to gentle and train horses. Millan too, it was implied, uses methods that really work because he understands dogs' true nature.

The trouble is, Millan's beliefs about dogs' true nature were based on the old adversarial story of relationships—that in the eternal competition for food and mates, those who succeed are higher in the pecking order. It's the old win-lose model in which two individuals are always in a one-up, one-down relationship to each other—the Malthusian story, filtered through Darwin, still alive and well in the twenty-first century.

And it's a story that leads logically toward harsh rather than gentle methods of animal training, for those who believe they have to remain in control are always tempted to escalate the amount of force they use. In fact, Millan has often been criticized by other dog trainers for his sometimes harsh and punitive techniques. One animal behaviorist, Nicholas Dodman at the college of veterinary medicine at Tufts University, said in 2006 that his school had written to the National Geographic Channel to tell them that by airing Millan's program, "they have put dog training back twenty years."

What Millan apparently didn't know was that by the time he started his show, the dominance story was losing ground in the very animal-science community where it had once flourished. The term *pecking order*, after being coined in 1922 by a Norwegian grad student studying the domestic chickens in his backyard, had spread like wildfire in the 1930s throughout animal studies. Only later was the question raised: Can behavior found among chickens be generalized to wolves and primates? Classic studies of wolves done in the forties discovered a pecking order among them, but years later the story came out that the studies were done on wolves in captivity, which one commentator notes would be like inferring human behavior from people in refugee camps. An even more apt comparison might be with prisoners—people forcibly separated from their families and thrown together, in confinement and under surveillance, with other displaced persons. Which leads me to wonder if pecking orders even among chickens are also the result of captivity—a pattern found only in flocks or herds of unrelated animals and exhibited only under stress.

Not to mention that after dominance had been used to explain everything from gender relations among primates, including humans, to wolf feeding habits, those who looked more closely found that the concept shrouded a dozen different meanings. Some researchers thought it meant animals controlling food access, while others used it to refer to bullying behavior, still others to the pecking order. And how to measure dominance? In aggression and fighting or in leadership qualities? Was an animal who adapted to another's wishes being submissive or merely respectful? The very concepts of dominance and submission, which for decades had seemed so clear, suddenly looked rather fuzzy. By 2009 behavior researchers studying human animals found that individuals increase their status in a group not by bullying or "pecking" but by making themselves indispensable to others—caring for and being generous to others.

There was also the issue, in science, of who was doing the observing. Scientists, like everyone else, are predisposed—by personal experience, by family history, by living in the Sonora Desert or on the Lower East Side—to see some things and not others. Even the gender of the researchers makes a difference. Twenty years ago two social scientists analyzed nearly a hundred articles in prestigious scientific journals written by primatologists in East Africa. They found that the women and men researchers tended to use different vocabularies to describe animal behavior. Women primatologists used the word *female* and related words (like *mother*) more frequently than male scientists. They also talked more about cooperation and its related words (*bond, affiliation, connection*), while their male colleagues talked more often of competition (and related words: *agonistic, aggressive*). The researchers were all watching primates in the same region of Africa, but they had brought different lenses to their work and so saw different things.

Like the primatologists, the American public brings its preconceptions to the matter of dog training. Despite the rethinking of dominance theories in recent decades, in the American mind the myth of the alpha dog lives on.

↶

IF WOLVES LIVE in families, not packs, then it makes about as much sense to speak of dominance toward your dog as it does toward your children. And yet until recently, dominating children is exactly what parenting in this culture was thought to be about.

I remember something my mother said—one of the many things she should not have confided to a daughter who was only nine: "We started too late with your brother," she told me one day. My brother, five years older, was "getting rebellious," as it was called in the sixties. "We didn't discipline him soon enough," she said in a soft voice, her eyes downcast. She was confessing. "We let him do what he wanted until he was four or five years old." She shook her head and, looking up, locked her eyes into mine. "You have to break the will of a child, you know," she said sadly.

Recently my fourteen-year-old brother had turned sullen and angry, stomping to his room and slamming the door, never stopping to talk to me after school or for that matter any time at all. The Bro I'd played football with and flown kites with, who'd taught me to ride a bicycle and shoot baskets, and whom I had generally worshipped for most of my life—even after he showed me how to fry ants through a magnifying glass—seemed lost to me. I felt shunned. I didn't know what to make of his new need for privacy. I hadn't yet heard of puberty.

My parents didn't know what to make of it either. In the generational conflicts of the sixties, they had several strikes against them: not only were they over thirty, they were nearly twice that age, as old as the grandparents of other teens. They were also deeply unhappy. My mother, sick and depressed, her anger increasing, had already gone on several rage binges. My father, with one polio-crippled leg, was hardly the portrait of dominant manhood; faced with life's challenges—more than the usual amount because of his disability—he tended to either feel helpless or turn rigid and insist on his own opinion.

In the face of my brother's fury, my father turned rigid and my mother went on a rampage. They searched every crevice of my brother's room, certain they would find drugs, a new problem that the high school had started educating parents about. What they found instead were girlie magazines stuffed under the mattress. I knew my brother had them—he'd shown them to me once when our parents were out seeing yet another of my mother's doctors, probably the same day he'd gone out to the garage to smoke. The pictures had raised mysterious feelings in me—I wanted to be like the women, all busty and bold and rebellious, none of which I was. But I hadn't felt comfortable being around my brother when he brought them out. He'd gotten a funny glint in his eye.

Now my parents, frantic, declared war on their son, my mother marching on him with threats and fury while my brother defended with stomping and shouting. I cowered in my room, listening to each battle through a closed door, trying to read or, if late at night, to sleep. Occasionally I heard a loud crash—was it just a chair against a wall? Or my brother getting slugged? Beating happened occasionally in the house; when we were small each of us had been whipped a few times with a leather belt or a palm on a bare backside.

As the family war escalated, my brother's behavior took a nosedive. He went out drinking with friends, coming home and puking in the bathroom. His grades sank. He took the family car out and wrecked it at midnight in a country ditch. During one argument with Mom he slammed his fist into the wall, leaving the imprint of knuckles in painted plaster.

He was desperate to get out of the house. Finally he negotiated a deal to enroll in a church-sponsored school in the next state for his senior year, boarding there with a local family. That year he got religion. He repented—I had no idea of what—and his faith took a rigid turn. His rage at Mom and disappointment in Dad grew a foundation of dogma: the Christian man was to be dominant in the family, the Christian wife submissive.

From that time on, my brother was a sober, upright, and fun-loving model of responsibility, cajoling his friends into karaoke, offering up an unending stream of jokes and one-liners (most of them off-color), working a white-collar job, and leading music in his church. But he remained volatile, as his family life had prepared him to be. The trauma of his early years drove him, as trauma always does, to reenact the painful scenes, desperately trying to heal them. Over and over he returned to porn, not understanding it was connected to his teenage pain. In secret he could relive the fourteen-year-old's frantic cry for help, but in secret he could not heal the rupture with his mom or win the respect and love he had missed out on.

After our parents were dead and Bro suffered his first divorce (a year after mine), his annual winter depressions turned more serious, and he became suicidal. While I was still battling melancholy, the year I was welcoming Sapphire into my Oakland apartment, he was crashing two thousand miles away in a psych ward in Ohio.

Even after he recovered, even after behavior treatment, he was quick to anger, and often his rage was directed at women. Nearly a decade later, now a truck-driving man of fifty, he called me one day from the cab of his semi, furious at his third wife for not following his wishes. He shouted doggedly, "The man must be head of his house! What do you have when two people are trying to be head?" I was about to offer, hopefully, "Partnership?" but before I could even form the word he had thundered on: "A two-headed monster, that's what!"

～

I TOO CHAFED under our parents' regime, though I was more docile than my brother and didn't receive the brunt of their anger as he did. Faced with an arbitrary ruling from our parents, I would slam a door, but not once did I get away with it. My father would limp up behind me and say in a soft, threatening voice, "Now open that door and shut it right." With every cell in me screaming, every muscle straining to slam it again and explode at the injustice, I would turn the knob, open the door, and latch it gently. And remain silent. I can only say that repression failed to give me tools for dealing with either anger or injustice.

In the calculus of dominance, it was clear, parents always came out on top. Family life boiled down to a battle of wills.

By the age of twelve I realized I would have to bide my time—go underground—until I could get out from under their roof. The other options were unthinkable: going insane with rage, like my brother, or leaving home and trying to support myself. Our family was so isolated and family violence so taboo that it never even occurred to me to approach other adults for help. And I was way too timid to imagine moving away, as my brother had, to live with a new family.

Besides, I had no faith that family life was different anywhere else. In my friends' houses similar scenes, and worse, took place.

The deeper problem was that my parents were hardly alone in trying to dominate their children. Getting children to submit to authority had long been *the* definition of successful parenting among my ancestors. I recently picked up psychologist Alice Miller's book *For Your Own Good*, a critique of the harsh European history of child rearing, and was horrified to find these gems of handed-down wisdom from old German parenting manuals:

> If one gives in to [children's] willfulness once, the second time it will be more pronounced and more difficult to drive out. . . . If their wills can be broken at this time [the early years], they will never remember afterwards that they had a will. (1748)

> Obedience is so important that all education is actually nothing other than learning how to obey. (1748)

Anyone who believes he can win love only if he is obeyed as a result of explanations is sorely mistaken, for he fails to recognize the nature of the child and his need to submit to someone stronger than himself. (1852)

A hundred and fifty years later, the same argument would be used with dogs. It's their nature, it would be said. They need this kind of submission.

Disobedience amounts to a declaration of war against you. Your son is trying to usurp your authority, and you are justified in answering force with force. The blows you administer should not be merely playful ones but should convince him that you are his master. (1752)

Children must first be trained before they can be taught. . . . Healthy discipline must always include corporal punishment. (1887)

Old habits die hard. In spite of Benjamin Spock and his innovative advice in the fifties to treat children firmly but kindly, a national survey done in the midsixties, at the very time my family life was erupting, showed near-unanimous support among American parents for using corporal punishment (94 percent). A few years later the popularity of spanking began to drop dramatically, yet the practice changed little; thirty years later, a 1995 Gallup survey of parents revealed that 94 percent of them still spanked toddlers. Centuries of culture are not unlearned in a single generation.

Alice Miller, a German Jew, believed the "poisonous pedagogy" of dominance and submission laid the foundation for fascism in Germany. Children raised in this way, she said, become capable of obeying even the vilest commands. The commandant at Auschwitz, Rudolf Höss, while in prison was ordered to compose a memoir, an order that he of course followed conscientiously. He wrote about his childhood, "It was constantly impressed upon me in forceful terms that I must obey promptly the wishes and commands of my parents, teachers, and priests, and indeed of all adults, including servants, and that nothing must distract me from this duty. Whatever they said was always right." Such obedience, he wrote, "became second nature to me." Miller adds, "The men and women who carried out 'the final solution' did not let their feelings stand in their way for the simple reason that they had been raised from infancy not to have any feelings of their own but to experience their parents' wishes as their own."

Hitler himself grew up in a tyrannical household and was beaten regularly—some say daily—by an alcoholic father. As a boy, Adolf loved to read adventure stories, and from their romantic tales of bravery he learned to take the beatings stoically. He even prided himself on counting the lashes. Is it any wonder that the grown man, newly installed as Führer of the Third Reich, suffered nightmares so violent he would wake up wild-eyed and shrieking, trembling so hard that the bed vibrated, shouting, "He's been here!" and rattling off what sounded like meaningless numbers?

It is a long way from National Socialism to the idea that animals organize themselves into pecking orders. And yet more than historical coincidence links them in the 1930s. What they shared was a decade of upheaval—political turmoil in Germany and economic depression in the States—as well as a bedrock practice, passed through families over centuries, of organizing relationships through dominance and submission. Though Americans of the time liked to think they would never give a political leader the kind of obedience that the German people were giving Hitler, they were more than willing to "find" a similar pattern in the natural world, to "discover" that violent force was ubiquitous among animals. It wasn't the pattern of dominance and submission itself that they worried about—after all, it was only natural—but rather that civilized Europeans were sinking to the level of brutes.

To this day German ancestry is shared by the largest portion of Americans; it remains the majority in the twenty-three states spread across the northern half of the country in a broad swath from Pennsylvania to the Pacific. And child-rearing methods of beating and parental dominance are not the only repressive legacy we as a country have inherited; the brutalities of African slaveholding and Indian genocide reverberate through the centuries, leaving marks of trauma on each succeeding generation.

Which is all to say that dominance and submission as a "normal" part of life, rooted in nature, were deeply embedded in the American imagination long before Cesar Millan began training dogs. To enjoy unprecedented success, he had only to tap into it.

⌒

AS IT HAPPENS, while I am writing this chapter I get a disturbing refresher course in dominating children. I am hiking one morning, leading a small group in a nature meditation as we climb a steep mountain trail close to town. The early-autumn day promises to be hot, but in the shadow of the hillside all is cool and quiet, the ponderosa forest calm but for the rustles of a spotted towhee in the brush. Partway up the long hill, the group disperses from the oversize rocky steps, each person seeking a quiet spot to dwell for a few moments and listen to the voices of earth and wind. I find a friendly, flat-topped rock and climb upon it, only to discover that I am still so close to the trail I can see and hear the other hikers.

A blond-haired couple and their young daughter are climbing the stairs. The little girl, no more than five, grabs her dad's hand to steady herself. She is mumbling in a low whine. I imagine she is tired, wants to be done with this hike. Her young father is getting frustrated. "Now, Alma, what is wrong with you?" he complains. "I know you, and you've done this kind of thing before. You're just putting this on. What is the matter with you?" A few steps later: "Alma, if you think we're going to carry you,

you're sadly mistaken because you're *way* past that stage. What is *wrong* with you, anyway?"

His voice fades for a moment, then: "Alma, stop grabbing my hand! It's slowing us down!" They round a bend, and the air quiets again to a woodsy peace.

But a few minutes later, from far above, a scream pierces the air. It is full-bodied and frantic, the scream of a desperately frustrated child. Then a whining wail and another shriek.

I want to rush up the hill, to say in a low voice to Alma's father, "She's really small; I bet she's tired." But how welcome is the interference of well-meaning strangers? So all I do is cheer Alma on in my mind.

Go, Alma! I silently shout. *Scream, child! Don't give up your voice!*

Alma's father probably never heard of breaking a child's will. But he is living out a modern variant: the role of drill sergeant, driving his children as hard as he probably drives himself—no patience for slowing down, no capacity for listening to the needs of his own small child let alone those of the trees around him and the land on which he walks. In the battle of wills he has set up in his family, he is determined to conquer. Alma's dissident voice will get no hearing.

~

DOMINATION. THE FUEL of death camps and of the mistreatment and abuse of children. But the engine too of less volatile events: the creep of a shopping mall across marshgrass, the choosing of deep-sea sonar over whalesong. The assumption that human voices count above all others, that nature must defer to human will.

I am deliberately mixing *domination* and *dominance* here because they share the element of coercion, a fixed inequality. A practice of pressuring others to bend to one's wishes. A relationship in which one side defers by fiat to the other.

Domination seems too long a word for something so easy to do, something done every day with habits and attention. Erasing the other. Erasing awareness of an other. Acting as if there is only one will, not two or more, to take account of. Alma screams to remind her parents that her voice too needs to count. But who reminds the human race of the needs of trees? Or that our health depends on the health of water and land? The screams of Earth are often more subtle—a gasp of grief in human chests at the sight of a strip mine, the ache of remembrance over a forest cleared. Even nature's loudest voices, like the roars of calving polar ice, have not yet fully captured our attention.

The alternative to dominance is partnership, and that's what I wanted with Sapphire. But to find it I had to give up the idea of controlling her. Not dominance but dancing is what I hoped for. Even that is not quite right, for in some dancing the roles are scripted and the man always leads. What I yearned for was more like jazz—first one of us then the other breaking out in a solo, and the rest of the time jamming playfully together on the same beat.

ᕲ

IT WAS WHAT I wanted with a man too, though I wasn't having much luck in finding it. When Sapphire came to live with me, I had just begun to date again, and soon it became clear that Sapphire would help me in my search. She was already bringing new people into my life—the writer whose ranch we visited to consult on a book; the animal communicator, Winterhawk, who, with her silvery butch, tiny braided pigtail, and merry brown eyes, soon became a friendly staple in my life.

In our early months together, I daydreamed that Sapphire would lead me to the love of my life. I'd meet him in the park, where she and I were walking, and the only reason we'd cross his path that day would be that Sapphire had been staring at my back earlier than usual, impatient to head out the door. Or I'd find some kind vet—alternative medicine, of course—and while he was examining her we would banter lightly and later go out for coffee then dinner. (I ran across several fine alternative-medicine vets, but the male ones were never available.)

Only a decade later, as Sapphire's life was drawing to a close, did I begin to understand what she was teaching me. She did indeed lead me to the love of my life, but not in the way I had planned. She led instead by example. She taught me intimacy—being attuned to each other by day, sleeping bumped up to each other by night. Enjoying the sight of the other across the room, and being enjoyed in the same way. Romping together through grass or shallow surf, delighting in another and receiving delight in return, the days a quiet hum of pleasure in another's presence. Through Sapphire I came to know that equality means mutual gladness; through her I came to know that it can be sustained over months and then years; and having gotten to know it with her—having learned to enjoy and be enjoyed—eventually I could not be tempted by less.

But getting there took some time. Until then, Sapphire became my guide. She escorted me across a threshold visible only in hindsight—a before of anonymity, an after of intimacy, as Celtic writer John O'Donohue puts it. I was stronger, happier, because she was there. She didn't replace the human connection I craved—my loneliness still bit sharply—but she was preparing the way. Through Sapphire, I was getting acquainted with love.

༄

WHEN IT CAME to men, I soon learned that Sapphire and I didn't always see eye to eye. Just before Sapphire joined me I'd placed an ad in the local free rag (online dating was still a few years away) seeking a man who was "self-aware, warmhearted, and a wilderness lover." Hoping for some rides on the back of a sport bike—a Honda or BMW or, if my luck was exceptional, a Ducati—I'd added, "Motorcycle a plus."

Of the men who responded, the best prospect was a soft-spoken, orange-haired Honda rider going for a degree in environmental studies. My favorite moments with him—aside from some warmhearted sex—happened on his motorcycle during freeway backups. Stalled behind a horizon of cars, he'd call over his shoulder, "Ready?" then flip down both rearview mirrors. With a twist of the throttle, we'd shoot smoothly forward between cars lined up just inches from the handlebars. At the head of the line of cars the mirrors would flip back up and we'd zoom away.

After Sapphire entered my life, Honda Man fell in love with her as I had. He became so happy to see her that whenever he walked into our apartment he couldn't resist picking her up, all forty-five pounds of her, as if she were a cat. She hated to have her feet leave the ground, and she'd fix her gaze on me, ears drooping—*Can't you do something?*—until I finally asked him to respect her wishes and stop lifting her up.

Sapphire's favorite turned out to be a singer-songwriter for children, a gentle, rounded man of thirty-nine with a shock of wavy gray hair. He was also a die-hard supporter of People for the Ethical Treatment of Animals, a detail I didn't find out until later but Sapphire apparently knew instantly. On the day I met Silver Teddy at a local café, we liked each other well enough to go on a hike together after coffee, stopping at my apartment so I could change shoes. I disappeared into my bedroom for less than three minutes and emerged to find Sapphire lying on the sofa wrapped in the arms of this stranger. Their eyes were closed, as if they'd been cuddling for hours. Sapphire turned her head to glance at me, and I swear I saw guilty pleasure on her face. I agreed that Teddy was huggable, but it wasn't quite enough for me, though we did remain friends.

⌒

THE LONGER WE lived together, the more Sapphire gentled me. It happened as I stopped to listen for her voice or paid attention to her perspective. That first time Winterhawk consulted with her over the phone, I discovered a lump in my throat through most of the conversation. I was catching a glimpse of the world through a dog's eyes, and it softened me. I've felt something similar while cradling a child—that flush of wonder at the sheer vastness of the being in my arms, a glimpse that brings tears to my eyes. But dogs can take care of themselves in ways that children can't, and so our relationships with our dogs tend to be a little more matter-of-fact, bounded by the knowledge that we dwell in completely different worlds—they relying on nose and ears, we on eyes and hands and intellect. We are not like our dogs, and we know it.

The knowledge can lead us to feel more separate than is necessary. Because I enjoyed the sense of communion with Sapphire, the softening, I sought it again. I began to experiment with sending pictures to her and receiving them in return. Each time I received a picture or a feeling that surprised me, I held open the possibility that it came not from me but from her. It was a meditation focused on image and feeling—a practice, I discovered, that not only fostered inner softening but required it from the start. A special gentleness was needed to begin imagining the world from a dog's point of view. I recently heard a grandfather speak of holding his hands open for his new grandchild. "Healing has something to do with holding a baby," he said. "It needs the preciousness of the touch we give to a newborn." My heart needed to be that tender, that open, to receive Sapphire's perspective.

Each time I succeeded in becoming that gentle, I was moved, often to tears. I found in Sapphire someone whose feelings and reactions were often similar to mine. And why wouldn't they be, if we shared a mammal physiology and a limbic, emotional brain? But her emotions were centered on far different concerns. Sapphire cared about things at ground level. She paid attention to squirrels—of course—and followed the doings of mice who crept through our apartment on winter nights. It was her job to manage happenings on the floor. Humans could take care of the stuff higher up.

Then there was that nose. I will never forget the day I meditated with her, asking her to show me what she experienced in the park that was so compelling that she had to run off the trail to follow it. As soon as I asked the question, my face flooded with a prickly sensation like wasabi up my nose but enticing, not painful, a mouthwatering smell like stew simmering on the stove. My body nearly jumped up out of the chair to follow it. If to her every scent was as compelling as wasabi and as appetizing as dinner cooking, no wonder she plunged away into the thicket.

One day several years after Sapphire joined me, while I was at work at my desk and she was asleep on the deck outside, I was startled to hear her barking. She was normally quiet—only one or two barks for the doorbell, none at all for the mail carrier—so when the barking continued, urgent and stern, I stepped onto the deck to investigate. I found a man climbing down the hillside between our house and the neighbor's. He looked up, nervous. He had expected no one at home.

"Can I help you?" I asked icily. Smartly printed shirt and jeans, in his thirties—he looked familiar. Had I seen him doing deliveries for a courier? He mumbled the name of someone he said he was looking for—no one I recognized—then headed quickly downhill toward the street. I raced out to intercept him, but he had disappeared into the brushy hillside.

Some days later I called my dear friend Jackie in Los Angeles to tell her about the intruder. Our neighborhood had started being the target of daytime burglaries, and I was sure he was scoping the houses for entrances and exits. "I wonder if Sapphire recognized him," I mused to Jackie.

"Why don't I ask her?" Jackie replied. She too was experimenting with sending pictures to her dogs and receiving them in return. She grew silent for a few moments on the other end of the line. When she began speaking again, her voice was hesitant, not believing.

"Sapphire says it was the other animals on the property who told her there was an intruder," she said softly. "She shows me pictures of all of them—the rats, the opossums, the squirrels. She knows where the snakes have their holes. She says it's her job to keep track of them all. And they— this is the part I can't believe—they too know who is and isn't supposed to be on the property. She says the ground animals woke her up and told her someone was there. That's how she found him, she said."

I listened, not believing. Whoever heard of such a thing—that the animals know who belongs on the grounds?

Then an old remembered phrase flowed through my mind: "So great a cloud of witnesses." In the New Testament I was raised to read daily, the writer to the Hebrews imagines a great throng of witnesses surrounding us, cheering us on like spectators at a stadium. In the days of the Roman Empire, the allies were always imagined as "up there," ensconced in the heavens, peering down at the mortals below. But Sapphire was giving us a glimpse of something different: they're not up there in the clouds; they're down below our feet, sharing the ground we walk on, the ground we think we own, the ground that gives us sustenance.

Our allies are right here on Earth.

~

ON A SUNNY but windy January day at Limantour Beach north of the city, Sapphire and I were plowing together through the dunes when I heard a faint whooping and shouting. I looked up to see, far away, the few other humans on the beach jumping and pointing toward the bay. There, in the shallow waters just yards from shore, was a whale. No! Two whales—one much smaller than the other. I'd completely forgotten it was migrating season; this mother and her young one were playing in the shallows.

Sapphire, far ahead, was keeping an eye on me that day rather than running away, so I was free to stand and watch for a long while. The whales swam back and forth, one moment beside each other, the next moving in opposite directions, their blowholes occasional volcanoes of spray.

The mother opened her enormous jaws, and the roof of her mouth lifted far above the surf line to reveal the cavern of her throat. Then the great jaw swung shut and she disappeared again under the surface. Eventually I lost track of them as they made their ambling way along the shore, and the cold wind forced me to call Sapphire and turn reluctantly toward the car. But I carried away from the beach the memory of those enormous, graceful bodies playing, and a quiet sense of blessing.

On a different day Sapphire and I were again at Point Reyes, this time on North Beach. As usual, Sapphire was trotting ahead of me, sniffing left and right to see if this would be her lucky day—some dead fish or rotting bird to roll in. A stiff breeze was blowing from our backs out to sea.

Suddenly I had the feeling of being watched, a presence just behind and to the right. I whirled in time to see a crow sail past my shoulder no farther than two or three feet away. The instant it passed, it pulled in its wings and flipped onto its back, then spread its wings for a moment and glided upside down. Without losing altitude it pulled its wings in again and rolled over to right itself, then sailed out to sea. Strong winds are known to flip birds over, but this movement was practiced, smooth, not a shred of flailing.

A crow showing off! I grinned. I'd just been treated to a private performance.

A sense of wonder lit the rest of my day.

For these gifts, I had Sapphire to thank. Without her, I'd have been tempted to stay indoors, to press on with editing or dissertation. But because of her I had to make time for play. I wasn't bird-watching as often anymore—binoculars jiggle when a curious dog is on the other end of a leash—but what she was giving me more than made up for it.

Sapphire was giving me partnership. She'd turned my apartment into a home. A family now lived there, a family of two. These two, though different species, loved each other and teased each other, hiked in woods and on beaches together, checked in with each other silently across a room, and took turns deferring to each other. I made her take baths once in a while; she made me take walks and play every day.

Clearly, I was getting the better deal.

7

Ancestors

When we love the earth, we are able to love ourselves more fully.
I believe this. The ancestors taught me it was so.

—BELL HOOKS, *SISTERS OF THE YAM*

MIMI AND I climbed out of my little hatchback and, suddenly giddy, turned in circles to take in the view. Immense granite boulders gleamed silver-white beside pine trees draped in plush green. We staggered as if drunk in the thin air. At seven thousand feet in the High Sierra, we had to be near the top of the world! Giggling, we hauled out our camping gear then, before setting it up, headed down the nearest trail. Too lightheaded to enjoy the walk and now sleepy as if drugged, we crept off the trail and lay on our backs in a meadow to gaze up at wildflowers nodding over our heads. Our eyelids soon closed.

When we awoke, the tipsy feeling had passed, and we could drink in our surroundings. The view went on for miles, down toward the Sacramento Valley below. A yellow haze filled the air from the normal late-summer wildfires burning elsewhere in the mountains. But here the granite boulders sparkled in the sunlight, and the turf lay thick below us, fresh with flowers.

It was August 1996, a year after meeting the eagle and nearly a year since Sapphire had come to join me. Though I'd gone off the antidepressant and was functioning again as if normal, I still did not feel internally well. Like Sisyphus, I woke every morning to roll the same mountain-size boulder up the hill. That boulder was despair. Not depression, exactly, but despair.

How do I characterize my particular boulder of despair? It was the feeling that nothing would ever change. That too many bad things happened in the world, and I couldn't do anything about them. That others magically knew how to be happy but that I was born without that knowledge, or that magic.

Now, fifteen years later, I see that my long drought of despair was a stage of grief—the protracted time of gray dullness after the shock of loss has worn off, after anger and bargaining have run their useless courses. For me that stage of grief is usually the longest, settling in like snow in a high valley that won't melt until summer. If in my midthirties the chill of death cast a longer-than-usual shadow, it's because the mountain of losses during that time loomed unusually large.

I may have felt despairing on the inside, but on the outside my life was improving. The end of the dissertation was in sight—maybe six months away. More than one night I dreamed that I had finally started my last year of school. Day by day that dream was coming true.

Yet the heaviness would not lift. Below the moments of pleasure I could always sense the pit, even while camping with my dear friend Mimi. In the sparkling Sierra sunlight after breakfast, as I scrubbed black and sulfured remains of scrambled eggs from the fry pan, I rehashed to her where I felt stuck: I had been born with a deficit—to parents who had passed their own crippled emotional lives down to their children, who had tried to remake us in their image, and whose guidance I had learned, long before I was a teenager, not to trust. Now they had been dead for four years, and there was no hope of getting what I'd always needed. How could I over-come that kind of deficit?

Just two weeks before that camping trip, I'd heard about a new healer. Chuck was a medicine man and came highly recommended, but I was wary: too many white people were running off to the desert for sweat lodges and becoming medicine people overnight. Yet I was tired, way tired, of pushing that boulder.

I went ahead and made an appointment.

~

EVEN THOUGH HE'D told me on the phone that he was a white man from Winnipeg, I was surprised when I drove up in front of Chuck's suburban house and tidy lawn. He met me at the front door, with snowy hair and bright blue eyes that gleamed and twinkled when he grinned, which was often.

I had brought what he'd requested—some pipe tobacco "for the Grandfathers" and an offering of "whatever Spirit makes available," since medicine people are not allowed to charge. My few dollars were enough.

In his healing room I sank onto one of the pillows lining the wall, and Chuck began packing his long-stemmed pipe with tobacco. There was a cheerful calmness about him, and I felt soothed as he lit the pipe and the smoke wafted toward me. Chuck explained what would happen: the smoke would be our prayers going to the Creator, Chuck would sing a little song for me, and I would be healed.

I smiled. As if it was going to be that easy. I was open to the possibility, of course, but up to this point healing had meant a whole lot of hard work, and I doubted that now would be any different.

He switched off the table lamp and chanted for some minutes. When the light came on again, he was beaming. "You've already done a lot of the work," he said. "Almost all of the pieces are in place. It's like a pie with just one small piece missing." He grinned again, his eyes twinkling. "Come back in a month," he said, "when I'm back from vacation, and we'll work on that other piece." I was to wear a ribbon of red, tied around my forehead like a bandanna, for the next four nights. "The Grandfathers and Grandmothers come to help during that time," he explained. "Red is the color of the west. It's the color of change. The Grandfathers from the west will come and help you."

After the ceremony I tried to feel happy. Only one piece missing. Good news, I supposed, though I figured he was wrong about the size of the piece. On the way home I stopped at a fabric store and bought a yard of red satin ribbon, which I tied around my head for the next four nights. I had lots of dreams as usual, and in the morning the red ribbon still circled my head. At the end of the fourth night I set it aside in a drawer.

꙳

AFTER MY WEEKEND in the Sierras with Mimi, I returned to Chuck's house for a second appointment. Again he led me into his medicine room, and we took our seats. He tamped down fresh tobacco in the pipe, telling me all about his summer vacation. I listened politely. Then he turned off the lamp and lit the pipe. Embers flashed in its bowl, glittering bright in the darkness then dying into smoke. He began the opening prayers, and this time I caught an English word here and there: "Creator . . . water spirits . . . we pray for Priscilla . . . we thank you . . . "

Today my mind was jumpy and I could not calm it. I was hyperaware of every unfamiliar detail in this unfamiliar room. What was I doing in the pitch-dark, listening to a man I barely knew chant alien words while the room filled up with smoke? Was he saying anything at all, or was it just syllables?

I caught myself thinking and took a breath. Could I be present? Here, now, this moment?

The smoke was thick, the rhythm of the rattle in his hand noisy but soothing too. I wondered how Chuck had come to be a medicine man. What was his story?

Oh, no, thinking again.

I stopped and breathed, trying to focus only on the sounds. The chanting went on and on, the rattle with its tiny stones within it rustling like the dry leaves of a hundred cottonwoods. Here, now, in this moment, what might be available?

To my surprise, as I focused on being present, I felt a question form in my mind, as if posed by someone else: *Do you want to feel better?*

Well, duh, yes! I'd been trying for years, in many different ways, to feel better. But how? The problem was I didn't know how.

As soon as I'd had this thought, a follow-up question appeared: *Would you feel better if you knew how?*

For a long moment I searched my inner rooms: Was there any part of me that didn't want to feel better? I could find none. My answer was fully yes.

I don't know how to describe what happened next. It was as if some load on my shoulders tumbled to the ground. The weight I had been carrying fell away—as if, in the moment of not knowing how to make this thing happen, the very need to make it happen dropped away, not because I had done something, but because this something had already been done, by somebody else, for me.

The stuck feeling had been removed.

I double-checked. Was it really gone? Yes, gone. I couldn't find a trace. Only spaciousness remained.

I burst into tears. Shy about crying in front of a stranger, I sobbed as quietly as I could, shielded by the chanting, by darkness, and by smoke.

Finally the singing stopped and Chuck turned on the lamp. I wiped my eyes and nose with my hand and looked up slowly. He was taking apart his pipe. "Ah, very, very good," he said softly. "I saw you let it all go." He swished a cleaner through the pipe. "You were stuck in the past. It's all about letting go."

I looked at him, puzzled. He went on, "It's like the monkey in the forest who finds a jar of peanuts. He puts his hand in the jar, but his hand gets caught, and he can't get the peanuts. He has to let go of the jar first."

That old story. I resisted the urge to roll my eyes. What exactly did his story have to do with me?

"You were stuck in the past," he repeated. "You have to let go. You can't go back and fix things. That's what you did today. You stopped trying to fix the past."

Laying down the pipe, he added, grinning, "It's great, great! You're going to have a great life!"

I nodded a little uncertainly. Something had happened, but what? And would it make any real difference?

"Wear a white ribbon this time," he said. "White is for the north, for healing. Come back when you need to. And next time," he smiled, "I'll have a box of tissues."

I walked out into summer sun, blinking at the brightness. As I drove home I felt sober and still inside, as if I had traveled very far away and it was taking a long time to return. I stopped again at the fabric store and bought a yard of white ribbon, tying it around my head for the next four nights.

It didn't take long for the change to become palpable. The day after the ceremony I felt a kind of shy newness, like a doe who steps into a clearing that until now she had only skirted. The feeling grew into one of lightness and ease, as if life was truly simpler than I had thought. The lightness persisted for a week. When, after a month, the sense of cheerful ease was still present, I realized that this was happiness.

Some mending of my soul had taken place. At the age of thirty-nine, I had become lighthearted—for the first time in my life.

~

I HAVE TO say a word here about white people going to Native ceremonies. It's very controversial. Many Indians strongly oppose it. They ask, Wasn't stealing our land, our homes, enough? Now you want to rip off our religion too? Indians feel that white folks crash ceremonies in order to get power, and that power, to white people, most often means making everyone around you do as you say, and Indians have had more than enough of that for hundreds of years. They resent "Indian wannabes" who romanticize long-ago Indian cultures but never lift a finger to help out with the urgent problems that real Indians face right now: broken treaties; poisoned water and air left behind by mining companies; violations of sacred lands; the poverty and despair that travel down through generations after people have been conquered and displaced, their identity snuffed.

In recent decades it has become popular, even fashionable, for white people like me to consult with a shaman or a healer. Some come away telling stories of miracles. Others begin holding sweat lodges, charging exorbitant fees for their services—with sometimes tragic results, like those who died in a sweat lodge in Sedona in 2009 after paying upwards of ten thousand dollars for the privilege. To Indians, commercializing is the worst desecration of ceremony. Getting rich off the suffering of others is unimaginable when a healer's purpose is to serve the community.

I have even heard of white people who claim they are "more Indian" than genetically born Indians because these whites are living ecologically in ways they think Indians on reservations have lost. Thousands of groups of "Indian hobbyists" throughout the United States and Europe organize powwows, perform traditional Indian dances, and sell Indian-look-alike crafts. There are even stories of white people trying to patent the American Indian sweat lodge ceremony, "since native people were no longer performing it correctly."

Indians have a word for this: *stealing*. It's the twenty-first-century face of colonialism—yet another effort by white people to wipe out real Indians, this time by taking Indian ceremonies far away from real Indian communities, which means far from the lands that gave them birth.

I know enough of Chuck's story to know he's not a thief. He was given the gift—and responsibility—of doing pipe ceremonies not because he asked for it but because dreams and visions came to him that led him to seek help from Indian healers, and separate dreams and visions guided his Anishinaabe (Ojibwe) teachers in Manitoba to recognize and train him. In Ojibwe tradition this is the mark of authenticity. Then he spent years learning how to be a healer, years spent cleaning up his own life and abandoning the urge to self-promote. "That's all ego," Chuck says of advertising or charging money, and those who are motivated by the ego are working for themselves, not the Creator.

I recently heard a respected Ojibwe elder, Tobasonakwut Kinew, say publicly in a conference of Indians and whites that the days of secrecy about Indian ceremonies may be over. He met with the pope, who showed respect for the sacred medicine bag that Tobasonakwut held by ordering a table brought into the room so the bag would not rest on the floor. And the Catholic archbishop of Tobasonakwut's region recently attended the annual Sun Dance ceremony in person. "It is a major breakthrough," Tobasonakwut said. With the Catholic Church showing a newfound respect for Indian sacred teachings, he suggested, it may be time for Indians to speak openly; humanity needs what Indians can teach. "All the secrets that were taboo are no longer forbidden," he said. "We need to sort them out and see what belongs to human beings all over the world."

I also know that many other Indians don't agree with him.

ﮯ

I RECOGNIZE THE temptation of an Indian wannabe. For me it arises when I hear Indian colleagues talk about their ancestors. They feel close to them and speak in reverent tones about what they've learned from their elders. Inside me I notice a twinge, and the only name I have for it is envy. To be able to glance over your shoulder to the past and find a wise presence! It is something I have never felt, at least from parents and grandparents.

I remember a scene in the funeral home where my mother's body lay. During an interval between visitors, the kindhearted Mennonite pastor who lived a block away from our family came up to me with a stricken look on his face. "Your parents visited me when you were in college," he said. "You know they did that from time to time, usually late at night, when they were troubled about something.

"They'd gotten a letter from you, and they'd twisted it all around. They were worried you were doing drugs." I rolled my eyes—the thought of it: me, a girl too timid to try anything interesting let alone something dangerous. "I begged them," he went on. "I told them right out, 'Don't destroy your children!'" He shook his head, his eyes clouding.

Chuck of course was right. It was all about letting go of the past, not longing for something I'd never had—to be seen clearly by my parents. When I let go of wishing for the past to be different, I got better. I healed.

But I didn't see at the time that a deeper ache was also at work. It was related to parents, yes, but parents in the larger sense of all those who came before—all the farmers and quilters and woodworkers and gardeners and seamstresses whose lives made possible my own. Go back ten generations, and they number over a thousand. Reach back only five more generations, say, to 1500, the time of Luther and Columbus, and without intermarriage they would number thirty thousand. It boggles the mind: five hundred years ago there were thousands of people alive whose genes would go directly into making my own. Thousands of grandmothers and grandfathers.

The problem is, I have trouble identifying with any of them. I don't pray to God as they did or believe "he" resides outside the Earth. I don't approve of clearing forests as my great-great-grandparents did or shunning those community members who make mistakes. In turn, the Amish-Mennonites among my ancestors would heartily disapprove of me—of the modern clothes I wear, of my "worldly" education, of my living "in sin" outside of marriage.

But it's not really about my Mennonite past. Thousands of ancestors who were not Mennonite also handed down a culture that is failing us. The habits they learned from their parents—the only culture they knew to hand down to their children—are the very ways that have brought us to, and perhaps catapulted us beyond, the brink of destruction. As adults we have uncovered the truth behind stories we were told in elementary school: that Columbus and Pizarro and Cortés and de Soto were not merely "exploring" and "discovering" new worlds but were out to plunder whatever riches they found to line their own and their monarchs' coffers. But we do not so easily see behind the myth of the frontier to remember that settlers too were destructive—single-mindedly decimating their new environment.

~

AROUND 1800, TWENTY-FIVE years after the founding of this country, the frontier had moved inland in each of the brand-new states. It was marked along its length by settlers mowing down the thick eastern forest with a ferocity that was legendary even as it was happening. European tourists documented the process, scandalized by such disregard for the land. The English had razed their own forests hundreds of years earlier so that now any remaining trees on their lands were precious. They were horrified at the clear-cutting—not unlike present-day Americans appalled at losing tropical rain forests hundreds of years after we've cut down 90 percent of the trees within our own borders.

The early settlers, wrote tourists, regarded trees as "a nuisance," which "ought to be destroyed by any, and every means." The man who could cut down the most was seen "as the most industrious citizen, and the one that is making the greatest improvements in the country." In the settlers' frenzy to rid the land of trees, they refused to leave even a few for shade around their houses. Forests were mowed down and left to rot; a Polish traveler lamented the "sorry sight" of unending miles of sawed and burned trees, "these gigantic cadavers, shorn of their bark and half-burnt, lying about wasted."

According to one English nobleman, the frontier settlers were "the most destructive race that ever disfigured and destroyed a beautiful and luxuriant country." This English lord took pride in managing his own lands for the long-term health of the soil, and he could not contain his disgust: "Fires are daily made to consume what the people are too idle or to ignorant to convey to a saw mill only a few miles distant," he wrote on a trip up the Hudson River. The "barbarous backwoodsman," with his "utter abhorrence for the works of creation," is intent on annihilating everything in his path.

In the first place he drives away or destroys the more humanized Savage the rightful proprietor of the soil; in the next place he thoughtlessly, and rapaciously exterminates all living animals, that can afford profit, or maintenance to man, he then extirpates the woods that cloath and ornament the country, . . . and finally he exhausts and wears out the soil, and with the devastation he has thus committed usually meets with his own ruin; for by this time he is reduced to his original poverty; and it is then left to him only to sally forth and seek on the frontiers, a new country which he may again devour.

In upstate New York the deforestation and mass slaughtering of animals meant that "scarce a living creature is to be seen." Formerly teeming with wildlife, the land "in the short space of 20 years [has] become still as death." The details of dragging bears from their dens for slaughter and competing to kill the greatest numbers of animals are gruesome. In a single day in 1807—so the referee who managed this hunting competition announced—two teams in Worcester, New York, slaughtered more than 2,300 squirrels as well as one porcupine and one bear. In 1820, in a similar hunt of birds, more than 5,300 were killed in one day.

At the end of his life, Judge William Cooper, the father of novelist James Fenimore Cooper, said proudly of the corn, wheat, and grass fields surrounding his New York town, "I look back with self complacency upon what I have done, and am proud at having been an instrument in reclaiming such large and fruitful tracts from the waste of the creation."

The European ancestors, to put it mildly, did not love what was here. The bounty of the forest, which had sustained whole populations of people for millennia, appeared to them a wilderness and a wasteland.

~

HOW CAN ONE explain such hatred of the land? The clearing and killing exceeded a drive for profit. Small animals were hunted in seasons when their fur was worth little; forests were felled and left to burn and rot. It was wanton destruction, almost a group frenzy to see who could annihilate the most life.

The environmental historian Alan Taylor says that the settlers had moved to the frontier from places where the wilderness had already been decimated—the Eastern Seaboard and, before that, England, Germany, the Low Countries. To them, a land without trees and wolves and bears and panthers was normal, indeed, inevitable. Destroying wildlands was what "civilized" people did. If the forest had to go—and surely it did—they would profit from its demise. They were lucky enough to be there to enjoy the plunder. Then what I think is his most important line: "Virtually no one envisioned a restraint that would preserve some of the wild plants and animals, because none of their forebears in previous settlements had restrained themselves."

It was an ancestor problem. The ways of life received from their elders taught them to destroy the life that was here. No one before them had coexisted with wild animals or dwelled in a landscape not rearranged by human hands. None had made a living from forests or made wild animals a staple of their diet. There were no precedents.

Well, how could there be? one might ask. The Europeans were agriculturalists. How can you fault the settlers for being farmers instead of hunter-gatherers?

But I think the settlers' hatred goes deeper than that. The job description of farmer does not usually include pillaging and plundering the Earth. Those who make their living from growing things may be more than usually committed to preserving life, to regarding the Earth's gifts as precious and using them sparingly. Had the settlers been clearing land only for crops, they would not have had to destroy whole ecosystems.

No, the settlers were looking to reproduce on this continent the landscape of their ancestors. What had been handed to them was a culture of clear-cutting, and by god, they would stay true to it no matter what.

Unlike Taylor, though, I don't pinpoint "lack of restraint" as the issue—as if people "naturally" are rapacious (shades of Hobbes and Malthus hovering nearby). It takes a special set of circumstances—market forces that encourage plunder—to make large-scale destruction attractive in the first place, and it takes a culture that disdains unimproved land to set the stage for it.

"Unimproved." As if the land in its natural state is inferior. The esteemed Judge Cooper called it "the waste of the creation," a phrase that echoes the "howling waste of wilderness" in which the Israelites wandered for forty years (Deuteronomy 32:10). The "howling wilderness" had long been a staple of Puritan literature; in 1666 the British Puritan Samuel Mather wrote, "This World is but a Wilderness, an howling Wilderness, full of Lyons and Leopards, Sins and Troubles." Throughout the frontier era, land in its natural state was not just inferior but an active enemy. "In the morality play of westward expansion," wrote environmental historian Roderick Nash, "wilderness was the villain, and the pioneer, as hero, relished its destruction."

It seems unimaginable to us now, so accustomed are we to thinking of wilderness—a place that looks untouched—as offering special, life-enhancing qualities. We drive hundreds, sometimes thousands, of miles to stand in a small piece of it, idealizing it because we have rendered it so rare. But the idea that land is wasted if not farmed is an idea with long roots in European history.

Thousands of years ago when the Germanic tribes were expanding throughout Europe, the lands where they settled were still heavily forested. Agricultural peoples, these Germanic tribes—the Angles, Saxons, Normans, Franks, who would later become Germans, English, Dutch—were seminomadic. They lived on a piece of land for a time then moved on. It was an unsustainable habit even then, though no one noticed at the time because land was plentiful.

According to Germanic culture, a family earned the right to occupy a plot of ground by clearing some trees, staking out fields, and planting a few crops. There was no concept of ownership, but by "working" the land a family proved occupancy. When a father died, he passed down to his sons the right to farm the land, and the early German word for "inheritance," *Arbi* or *Erbi*, is related to the word for "work," *Arbeit*. Land, to be occupied, had to be worked.

Which is why John Locke, the seventeenth-century English philosopher whose ideas came to define English-language land laws (and a lot of our political ideas), wrote, "As much land as a man tills, plants, improves, cultivates, and can use the product of, so much is his property." The idea was hardly original; Locke was merely setting down the age-old Germanic idea that land, to be considered yours, had to be cleared and planted.

⌒

LAND PRACTICES AMONG the Romans were even less sustainable, bolstered by attitudes of arrogance and the will to control nature. The statesman Cicero (106–43 BCE) boasted in the early days of the Roman Empire:

> We are the absolute masters of what the earth produces. We enjoy the mountains and the plains. The rivers and lakes are ours. We sow the seed, and plant the trees. We fertilize the earth. . . . We stop, direct, and turn the rivers: in short, by our hands we endeavor, by our various operations in this world, to make, as it were, another nature.

His all-too-modern-sounding words might have been a template for the American frontier. They remained the template for Roman attitudes toward land throughout the centuries of empire. Cicero's contemporary, the poet Lucretius, complained that it was a serious "defect" that so much land was "greedily possessed by mountains and the forests of beasts." Lucretius applauded the farmers who compelled forests to flee up the hillsides, for it freed more land for the refined pursuits of winemaking and olive growing. Agriculture was civilized; uncultivated wilderness was not.

Farming was civilized because it made possible a comfortable life in the city. Roman life, in Italy as well as the provinces, was organized around the town or city as the core, supported by surrounding fields. Productive agricultural lands were divided into huge estates owned by a tiny fraction of families, whose patriarchs became the governors and judges and lawmakers of the urban bureaucracy.

Romans had inherited their focus on the city from the Greeks before them. Socrates had bluntly said, "Trees and open country won't teach me anything, whereas men in the town do," and throughout the centuries of the Roman Empire urban life remained central. Land outside the cities was useful only to the extent that it produced food for the urban core. Land that was too hilly to cultivate or too full of wild animals was worthless, good only for raiding to provide the raw materials for city life.

With this urban-focused dichotomy between field and wilderness, Rome depleted its natural resources and undermined its economic base. Romans deforested their hillsides, cutting wood for fuel and buildings, for warming braziers and war-making. When still more wood was needed, they seized forested tracts from others and proceeded to cut them down too.

Clear-cutting was not the only threat to the land; ground cover in North Africa was lost to a plague of hungry goats. The lions who once roamed the North African wilderness were hunted to near extinction for "games" in Roman arenas, and, lacking large predators to keep them in check, wild goat populations exploded, stripping vegetation from the hillsides. Soil erosion followed, depleting the breadbasket of the empire and putting a dent in the society's ability to feed itself. By the time Christianity became the state religion in the late 300s, soil erosion around the Mediterranean was a serious problem, and environmental degradation and resulting rising prices contributed to the empire's demise.

ᕤ

THE FRENCH TRAVELER Alexis de Tocqueville in his 1831 trip to the Michigan Territory wrote, astonished, that though Europeans talked a lot about the American wild lands, "the Americans themselves never think about them. . . . Their eyes are fixed upon another sight, the . . . march across these wilds, draining swamps, turning the course of rivers, peopling solitudes, and subduing nature." Like the ancient Romans, they were intent on creating "another nature."

More than anything else, their actions announced in no uncertain terms, "This land belongs to us now." The European tide would try to wash all evidence of natives—human and other—from these shores. They would replace as much of it as they could with their own plants and animals, with themselves and their children.

Deforestation moved west with the settlers. By 1850 my home state of Ohio, which local legend says was so heavily wooded that a squirrel could travel from one end to the other without touching the ground, was almost barren of trees. During those decades of cutting and burning, smoke blanketed the state for months on end. When my German ancestors arrived near midcentury, the only remaining affordable land lay in the forbidding Great Black Swamp near Lake Erie in the northwest corner of the state, a wetland shining with black waters and thick with life of all kinds, including mosquitoes. My ancestors then helped to decimate one of the few deciduous wetlands in the whole northern hemisphere, clearing the area of trees and artificially draining the soil.

State and local governments in Ohio encouraged soil drainage throughout the rest of the century so that by 1900 what had been a thick tangle of subterranean tree roots stretching from one end of the state to the other was replaced by thousands of miles of drain tiles planted underground to carry away the excess water that the deforested land could no longer absorb. (By 1960, it is estimated, nearly a quarter-million miles of drain tiles had been laid in Ohio, or enough to reach from the Earth to the moon. In the last fifty years most of those clay tiles have been replaced with plastic.) In this way the settlers "reclaimed" untillable land and made it produce. Following in the footsteps of their ancestors, they saved "the waste of the creation."

Still the settlers marched west, destroying as they went. On the other side of the Mississippi, in the wide-stretching plains, they slaughtered the buffalo then plowed up the prairies and planted them with European grasses to feed their European cattle. Historian Elliott West says that these "soil-rippers" did more than any state or local government to seal the advance of white society. "This ecological conquest was more lasting than all military campaigns and any government policy," he writes, "and its chief agent, the most effective enemy of the Indians' pastoral economy and lifeways, was the pioneer family."

The pioneers were immigrants and conquerors. They had arrived at a place they could not love. They would seize a world that was new to them, destroy what was here, and repopulate it with their own kinds.

Chickasaw writer Linda Hogan, whose ancestors were herded along the Trail of Tears to a reservation in Oklahoma, says she will never understand such systematic destruction. "Our world changed from one where every place and thing mattered and was loved, into a world defoliated, where nothing, human or other, mattered."

↶

To love "every place and thing": it was not a value I inherited. Of course, people come to love their homes, and over time European Americans too became attached to the lands they grew up on. But to this day the love often consists more of knowing how to work a place than of knowing how a place works. As if we were enslaved to ancestral ways, most white people I know still work the land to get from it what we want—a good hike, a satisfying climbing experience, a green lawn, the maximum yield. Far less often do we stop to merely gaze at the land, open-eyed—to watch it, as students, over time. Most of us have only a vague idea of which animals or plants are native to our region or how they evolved together over eons. In cities or suburbs, where 80 percent of Americans live, people drift atop markerless seas of concrete and mowed grass, with no remnants of native landscape to tell them exactly where, on Earth, they are. Climate change leader Bill McKibben goes so far as to call suburbs, with their miles of clipped lawns, "a device for making sure you never notice the natural world."

Knowing where we are takes attention. Sustained looking. Seeing. Becoming rooted in the place we live, whether bright desert or blustery mountain, dark woodland or windy shore. "Place is not simply location," wrote Celtic spirituality teacher John O'Donohue. "A place is a profound individuality." Though many modern people express a longing for lost wildness, what I think we are really missing is personality—the integrity of place. A relationship with the unique piece of Earth where we dwell, an ability to see its deep character.

The opposite of love is not hate, though it is related; it is not seeing.

↝

IN THE YEARS since I visited Chuck's medicine room, I have learned other ways of communing with Earth, and most of them don't look religious. A favorite is tending native plants. This morning I join volunteers in a native-plant garden at a public park in Boulder. It is the height of fall, with golden leaves crackling underfoot and azure skies soaring deep and endless above. October is the time to gather seeds.

Armed with gloves and a clipper, I set to work on Porter's asters, a small mound of a plant that last month swam in a cloud of tiny white daisy-like flowers. Now its stems have turned brown, its blossoms dried into small crunchy bulbs. I maneuver carefully around the sharp spires of a yucca to clip the top few inches off all the stems. A piece of prickly pear dislodges onto my glove, its spines piercing my fingers.

The asters were happy this summer, and I pack three large brown paper bags with their abundance. Our crew leader, Dave Sutherland, drops by to collect the fluffy towers of dotted gayfeathers. Dave is a naturalist extraordinaire, and from him I have learned most of what I know about Colorado wildflowers. He leads vigorous, laughter-filled hikes along local trails, and I have filled pages with notes about plants whose histories I forget but Dave does not. Dave too has some Mennonite ancestors, though farther back in his past than mine. We have talked often of the weeds that stowed away in the grain bins of our ancestors on the boats to America— the very exotics we now pull so that native wildflowers can flourish.

We kneel side by side, clipping. The midmorning sun is warm, though a couple hours ago my thermometer read twenty-eight. I feel happy, as I always do when tending the native plants. It's a feeling that envelops me as predictably as water closing over a swimmer. Dive into the native flowers, and I can expect to feel content. Full. Grinningly happy.

"I've figured out why I love working with the native plants," I say. Dave waits. "Our ancestors spent a lot of years hating what was here, trying to replace it, kill it, make it go away. Now here we are, loving it," I say. "Loving what is here."

"I like that," Dave says. "Appreciating the bit of original beauty that remains." Treasuring individuality, working to protect the personality of this place. "We're gardening with our heritage," he adds.

We are reopening a path of love between the humans and the flora of the land. It has been centuries since the ground enjoyed this kind of recognition. We are cultivating our relationship with ancestors, awakening to our connections with the soil and plants who knew this place thousands and millions of years before us.

What we are doing also purifies, as does the smoke of the medicine room. This too is ceremony.

⁓

"THE MOST RADICAL thing you [white people] can do," says Oneida writer Leslie Gray, "is to start thinking of yourself as having come from someplace in this land. That thought alone is going to be a huge contradiction to the prevailing models." The cure for wannabeism is to become indigenous in the way that is available to all people: forging our own relationship with the land where we live. When people sink their roots into a place, they and the land become one, said theologian Vine Deloria Jr., and when a spiritual unity is cultivated in this way, he said, the land is "consecrated." The Pawnee attorney and scholar Walter Echo-Hawk adds, "To adapt as the native people have done to the place where they live is to make peace with the people and the land."

Each of us can watch and learn—noting when each bird returns to our area in the spring. Tracking their habits with close attention, as we would the travels of our friends. Learning the tracks of animals, tracing their outlines in mud as lovingly as we might the whorls of a child's painting. Noting the moon phases, the sunrises and sunsets—not only when they happen, a minute or two or three different from the day before, but *how* they happen, each one unique, its personality shaped by cloud and wind and temperature. We can teach our children how to root themselves in place; we can be instrumental in training our various communities to do likewise.

I recently experienced a Maya elder explaining to a mostly white audience how to talk to the sun. "In the morning," he said, "when sunlight cracks your window and falls on your face, do you do this?" He reeled backward, pulling imaginary covers over his head, and muttered, "Oh, shit!

"Or do you do this?" He stood up straight, opening his arms wide to the imaginary window. His face aglow, he shouted, "Thank you! Thank you! Thank you!"

Cultivating relationship with land means being responsible to the creatures who live where we live, paying attention to their needs as if they were members of our own family. Which, of course, they are, each individual and species cocreating the landscape—the prairie dogs feeding eagles, the oak trees nourishing squirrels and deer with their acorns, the trees along the urban sidewalk refreshing the air. To the microbes in the soil: we can work to protect them from devastations of poison. To the fish: we can free streams and rivers so that waters can flow and fish can run. To the birds and animals: we can restore habitat so they will have places to migrate and to live. To the trees in the urban sidewalk: we can murmur, "Thank you, thank you." To the birds twittering in their branches: we can say, "Hello, my relative."

But the most basic gift of love is simply attention—taking time to notice how the sunlight lands on the tree outside the window at high noon in summer. Listening to the rustle of grasses in their spring delicacy and fall crispness, or distinguishing the whirring of wind in pines from wind in spruces. These are not skills modern people must depend on for our food—few of us hunt or farm for a living—but they remain the currency of friendship. Being delighted by the mystery of another and simply dwelling in that one's presence.

It's a close attention that is practiced even by some who would be reluctant to call it love. I think of a story that nature writer Kathleen Dean Moore tells about giving a speech at a ranger convention on the joys of loving the land. Afterward a park ranger from the audience came up to her. "I like what you say," he told her, "but I wonder if you couldn't say it without using the *L*-word." Moore was tempted to regard him as another casualty of what she calls "the sadness of all science," which trains its practitioners to become "like secret lovers," thrilling to the touch of the breeze or being astonished at the intricacy of genomes yet forbidden to speak of their feelings.

But then she reconsidered. "So what word," she asked the park ranger, "shall we use instead of l—e?"

She says he paused for a long moment before answering.

"Maybe, instead, we should say *listen to*."

Giving full attention to. Dwelling in the presence of the other. What if we truly listened to the world?

⌖

TO PARTICIPATE IN such an enormous cultural shift—to chart a new course based on listening instead of controlling—will involve rejecting, en masse, many ways of our human ancestors. Few of us seem willing to suffer this depth of alienation, to leave so much behind. Perhaps that explains the crippling inertia of our time, the stupor in which we keep slogging, robot-like, toward the abyss. We simply can't imagine that those who went before us could have been so wrong. But the path toward control stretches away from love, and too much of what Western society has received from ancestors leads not toward love but away from it.

The African American writer and theorist bell hooks credits the life-affirming ways of her ancestors, who loved the land, with giving her courage to heal the inner wounds inflicted by a society that dominates instead of listens. "When we love the earth, we are able to love ourselves more fully," she writes. "The ancestors taught me it was so." Healing the inner wounds opens people to experiencing more of life. "Like our ancestors using our powers to the fullest, we share the secrets of healing and come to know sustained joy."

Linda Hogan tells in her memoir of walking on a beach with a Chumash friend and encountering a loon who could not fly, for its feathers were tarred with oil. The loon had sunk into the sand, resigned to its death. The two women took the loon home and after many hours located someone in wildlife rehab who could help.

Afterward, the friend's father called Linda, in Chumash, the Girl Who Saved the Loon. "I tell this not to make myself look good," Linda writes, "but to reveal our values. This is what we love about our elders, that they honor us when we care, not when we win, but when we look after the earth and show compassion."

What will it take to become a society that praises those who care? That honors kindness more than success? That teaches children to love the Earth more than accumulate its products? I suspect we will need to listen to different elders—not the ones who promised wealth but the ones who taught compassion. We have such ancestors in Western culture, though their voices are often drowned out, and often by the very people who claim to be their followers. "Blessed are those who show mercy," said Jesus in the Sermon on the Mount, a very Jewish sermon in its concern for the poor and displaced. And later in the same sermon: "Ponder the wildflowers, how gloriously clothed they are—richer than the richest king."

These too are the values I was steeped in by my Mennonite forebears. This too is my heritage.

⌐

THINGS CHANGED RAPIDLY for me after that second ceremony with Chuck. My life soon filled with a new circle of friends, and I began enjoying hikes and poetry readings in the company of others. By Thanksgiving I was writing the final paragraphs on my dissertation—sooner than I'd expected—and the defense was to take place just after the New Year.

I continued driving to Chuck's house and attending ceremonies every week or two. Learning to live in the present, without dipping into that old valley of despair, turned out to be an ongoing practice. I certainly felt better day to day, like I'd suddenly been handed a brand-new well of reserves, but the work of forging new habits was ongoing. Each time I found my mood slipping, Chuck would repeat that I had gone again into the past, and he would do another ceremony, and I would wear another ribbon on my head and look forward to—or dread—the intense dreams and feelings that began to surface during those four nights.

Ceremony fed me. It soothed me to ask for holy help from the friends I was cultivating in the rest of my life: animals, water, land. In ceremony I could dwell for a time in dreams and visions, savoring connections between the outer world of nature and the deep inner places, drinking of the fresh springs feeding both. Once in a great while I could even feel the air in Chuck's room displaced by wings of helper spirits and hear the *whoosh-whoosh* of their beats. In ceremony the pictures of my heart could be seen clearly without my having to speak much about them, and I sighed with relief at being deeply known.

Most of all, ceremony was a time of being still, of opening to Earth and asking, "What now?" It was the other side of the healing coin: in therapy I was talking; in ceremony I was deeply listening.

～

ONE PIECE OF my life remained incomplete: I hadn't yet found a boyfriend. Just after Christmas, with time on my hands now that the dissertation had been handed in, I answered a personal ad listing all the usual things, like walks on the beach. After I'd left the message for this man, I wondered why. I mean, couldn't I have looked for something more original? Our first phone conversation left me a bit underwhelmed; we didn't seem to have a lot in common, and things would probably go nowhere. Still, over years of blind dating I'd learned that phone impressions could be misleading—for good or ill—so I arranged to meet him anyway for coffee.

The rest of that week I had other things to worry about. My dissertation defense was to take place on Friday. I reviewed my research, rehearsed how I might answer various questions, but felt unusually calm. It was another new trait; since that moment in Chuck's medicine room I didn't worry about things as much, as if I'd magically developed an ability to take life more in stride.

On Friday morning I sat through two grueling hours of conversation about my dissertation and then left the room while the faculty made their decision. Just after noon I received the happy news: the committee had passed my work. If I made a few revisions I would finally, after a dozen years, be done with graduate school. That evening I gathered new friends and old together to celebrate over chips and beer.

The next morning I left the apartment for the café wondering, as I always did, why in heaven's name I was doing this yet again. The men were never good matches: too wide, too thin, too cocky, too tentative. I reviewed in my head: the guy who, during our first (and last) phone call, asked me hesitantly if I would do "certain things" for my man, like wear stiletto heels, then hung up in a huff when I said that didn't do it for me. The guy in the coffee shop who put me through twenty questions in the first two minutes and, apparently hearing the wrong answers, said good-bye and left.

Yet here I was, heading down the road for another blind date. I walked into the coffee shop, and a good-looking man with a head of dark hair and a neatly trimmed goatee stood when I entered, smiled, and introduced himself as Anthony. I grinned broadly, breaking all rules of cool. When we parted a couple hours later, I knew I would see him again.

8

Peralta Creek

I came where the river
Ran over stones:
My ears knew
An early joy.

—THEODORE ROETHKE, "THE WAKING"

"**Y**OUR HOUSE IS here."
I was driving home from the park with Sapphire in the backseat. It was early spring 1998, just weeks after Anthony and I had moved in together to a rented flat. In the more than a year since we had met, we'd hiked the trails of the East Bay hills, gone camping together, enjoyed candlelit dinners with wine and chocolate, and, yes, gone for walks on the beach, with Sapphire running ahead joyfully to retrieve the sticks we threw for her into the shallow surf. The Pacific had been unusually warm that year, the only summer in decades that bodysurfing was possible, and that summer it was euphoric.

But then, the whole past year had been euphoric—my fortieth birthday in March 1997, commencement and a PhD in May, and falling in love slathered over the whole like icing on an ambrosial cake. Could life get any yummier?

Over the summer Anthony and I had settled in to getting to know each other, our steps at first tentative, then more sure as we came to trust the ground between us. By September we'd begun house hunting but after some months of no luck finally settled for renting. Yet even after we moved in together, we kept looking at houses for sale, though I didn't know why; I had zero interest in packing up again.

But our realtor seemed determined. "You'll price yourselves out of the market," she clucked at us one day, and it was true; home prices were surging. Every few days she would call with still another charming, older home to show us, and I'd grab my coat and umbrella, slip into still-wet shoes, and head out again into the rain. The warm ocean of the previous summer had swelled into the strongest El Niño on record, and this winter the rains had been torrential—day after day of downpours.

Now it was March, and Sapphire and I had taken advantage of a break in the weather to go hiking among pine and oak in Redwood Park at the crest of the hills. We were on our way home, coasting steeply downhill toward the tightest, creep-around-it hairpin turn in all of the Oakland hills, when I heard the voice in my ear.

"Your house is here," it said. The words were clear, as if someone had spoken them aloud. I checked the backseat; Sapphire was dozing, unaware. I braked and edged around the turn. Shocked, I listened for more. There was no more. Silence.

That evening my attention turned to other things—dinner, the movie Anthony and I were watching. I didn't mention the voice.

The next day Esther, from my old in-law apartment up in the hills, called to say I had mail there, and as soon as the rain let up I went to retrieve it. This was my first visit to the neighborhood in the more than two months since I'd moved out, and I'd been missing these overhanging bay trees and eucalyptus, the wide-open views of San Francisco Bay. Already the winding drive felt different, its curves not so automatic, the details of branches overhead no longer so familiar.

A block from my old apartment I negotiated a sharp bend in the road and came face-to-face with a FOR SALE sign. For three years I had driven this street almost every day, past intriguing-looking houses perched high on the steep slope above or hanging down off the road on the slope below, and in all that time the only property I was curious about, day after day, was the one on which this sign was planted. There was no house that I could see, only a beaten-down white garage backing into the steep hillside. But next to the garage a meandering flight of rough stone steps led up and up and up, disappearing into enormous bay trees above. Almost every time I cranked this turn in the road, I tried to follow those steps up toward where the house must be. Months before, I'd even pulled up to the battered garage and parked my car a moment to gaze skyward. I could see nothing but a few windows near a foundation, the rest of the house veiled in bay leaves. What kind of people lived there? Who climbed that towering tree-cloaked stairway each day just to reach their own door?

Better let our realtor know about this FOR SALE sign, I thought.

That weekend our realtor took us to two houses—the first one, near where we were now living, and this one, close to my old place. Secretly I hoped one of them would be ours, but really, what were the chances that a voice in my head would turn out to match reality? Still, I was looking forward to climbing those stone steps.

The first house was too close to the street, too dark, too much traffic. We drove then to my old winding street and parked in front of the rundown garage. Climbing out of the car, I heard the sound of rushing water and turned to see something I'd never noticed before: a creek running in a small bed alongside the garage. The steady rains had swelled the creek to a continuous surge, and its gurgling rush accompanied us as we began our long climb.

Halfway up the winding steps, I noticed my breath quickening—only partly because of the climb. The hillside felt magical. Lit by an overcast sky, the forest glowed. Emerald moss blanketed the boulders at our feet. Between the boulders lush ferns sprang out in astonishing green. Overhead, bay trees arched their enormous trunks, spreading a riot of leaves in radiant olive. The house was a cloud-hung castle above us, the creek marking a path at its feet. Step by enormous stone step, we were being lifted out of the noisy world below.

Finally, nearly ninety steps from the street, we reached the front door. Inside, another scene of magic—light pouring in, even on this cloudy day, and the San Francisco Bay a huge, murky jewel far below. The Golden Gate Bridge was a tiny silhouette far away, marking a spot on the horizon, I later learned, where the sun set twice a year—by coincidence, on the equinoxes, which meant on my birthday, the first day of spring.

Anthony and I trooped through the house. I loved everything I saw— the dated, dark redwood trim (charming!); the patio hanging out over the hillside (great view!); the three stairways (space separation!). The house was tall and thin, only a couple of rooms on each floor. In the living room I admired the small stained-glass window glowing amber and blue. On the next floor down, the kitchen sparkled with white cabinets and tiled counter. When Anthony and I finally met up in midgallop on one of the stairways, he asked with a smile, "Did you see the attic?" I located the attic stairs in what should have been a closet and headed up the steep steps. There I found the master bedroom, dazzling bright under a huge skylight, with dormer windows reaching westward from the roof to brush green treetops. I had just discovered my childhood dream—a treehouse—with a stunning view of the bay and the Golden Gate Bridge thrown in for good measure.

Wonder of it all, Anthony loved the house too. In the nearly six months we'd been house hunting, this was the only one both of us were enthusiastic about. The price of the house lay just beyond the ceiling we'd set—doesn't it always?—but even our second thoughts some days later matched our first: this house was truly ours.

Best of all, in addition to the view, was the creek. For on the day we first visited the house, the gray air was early-spring mild, the windows had been thrown open to the woods, and the faint rush of flowing water followed us from room to room. Spurred by El Niño rains, the tiny seasonal creek would flow that year until June.

I spent nearly six years in that house, years filled with adventures both heartbreaking and grand, but the most outstanding adventure of all involved the creek. And if I had to say who that voice in my head belonged to on the day it informed me our house was ready—at the same hour, I learned, when the FOR SALE sign was being hammered in next to that run-down garage—I would say it belonged to the creek.

⁓

OUR RELATIONSHIP WITH the creek started innocently enough. Before heading out of town to their new thirteen acres in the mountains, the sellers gave us some last-minute instructions. "Every year we do a little creek cleanup on the property," they said. A day or two before the city's annual large-item trash pickup day, they would hike upstream to pull out junk from the creekbed. "It's our little community service," they grinned. "We leave that job to you."

Trash pickup that year was scheduled for May, only a few weeks after we moved in, so late one afternoon after Anthony got home from work, we donned our oldest jeans, boots, and gloves, hiked the eighty-seven steps down to the garage, and picked our way carefully up the slurpy creekbed. Trickling water kept us company as we clawed our way around boulders, shrubs, and poison oak. We'd gone only a few feet before we found the first old tire lodged in muck. Then two more, then three. Before long we counted two dozen more. A few we could pry loose easily, and we rolled them out to the street. We lifted out a roll of rusty wire and moldering garden gloves, but the half-buried water heater and huge old storage tank would have to wait. This job was too big for two. We needed help.

And that is how I came to organize a creek cleanup.

⌒

"FOR MOST OF the year," the speaker was saying, "stream flows are highest at midnight and lowest at midday." I was attending a forestry conference a half-dozen years after meeting Peralta Creek, and the speaker was James Kirchner, then teaching earth sciences at UC Berkeley. He was talking about what can be learned about a forest from studying its streams.

Most of the year, he said, creeks keep up a predictable daily rhythm, surging in the middle of the night and subsiding at noon. They ebb and flow because trees and plants transpire, or breathe water out of their leaves and needles. When water evaporates from a tree's leaves, the lowered water pressure at the canopy causes the tree to siphon water upward from its roots. A large, mature tree may breathe a hundred gallons of water a day out of its leaves, which means it is drinking that much from the soil into its roots. Multiply those gallons by the thousands of trees and plants lining a creek, and the amount of water being sucked out of the ground approaches a staggering sum. Most of the drinking in takes place during daylight hours because moisture evaporates more rapidly from leaves when the air is warm. As the plants and trees suck water from the ground, less is available for the creek, and its flow subsides. At night, when transpiration slows, more water in the ground flows into and down the creek.

"Following the stream flow," Kirchner said, "gives us a way of watching the whole watershed breathing in and breathing out."

I blinked, astonished, not expecting to hear about breathing from an earth scientist.

Breathing in, breathing out. The most fundamental act of life—taking in energy, expelling waste. Inhaling oxygen, exhaling carbon dioxide. What mammals do through lungs, fish through gills, and insects through a complex system of tubes and vessels. What cells do across their cell walls. What anaerobic organisms do with sulfur or some other inorganic substance. What plants and trees do as well, using up to half of the oxygen they produce in photosynthesis to feed their own cells. And what they do with water too, breathing it in through their roots and out through their leaves (though sticklers say this is not breathing per se because it is merely transferring water from soil to air, not transforming water into energy).

Breathing in, breathing out. The rhythm upon which all living rides. The "fundamental activity of the universal being," according to Zen teacher Shunryu Suzuki. "What we call 'I' is just a swinging door which moves when we inhale and when we exhale." The most basic form of meditation is just to be aware of this simple, usually unconscious, act.

⌇

TO THE RARÁMURI people of the eastern Sierra Madre in Mexico, to live is to share breath with all others. *Iwígara* is the interconnection of all life, and *iwí* means to bind all together, like a lasso, a circle of life that the Rarámuri celebrate by dancing in ritual circles. But *iwí* can also mean "to inhale/exhale," and everything that breathes has a soul, an *iwí*. All who share breath, or soul, are kin.

The English language too remembers this age-old kinship of inner life and outer breath. *Respire* and *transpire* both come from the Latin *respirare*, "to breathe," also the root for *spirit*. *Respirare* itself is derived from the more ancient Indo-European **speis-*, "to blow," a syllable that like the word *hiss* enacts a breathing out.

Breathing in, breathing out. Respiring, transpiring, perspiring. Air and water circling, moving in and out, a flow, a rhythm, an endless undulation as long as breathing lasts. We step toward greater life in aspiring; we share breath when conspiring; and at the end of life, when the breath flows away one final time, we expire.

Breath is life. In-spiring.

To breathe is to create.

~

MONTH BY MONTH, I began to be inspired by the breathing, living creek. Gazing down on it from my kitchen window, I memorized the precipitous angle of its tall banks. I tuned my ear to the far-below gurgles, which tapered into silence as the creekflow dwindled then died during the summer dry season. The next winter, when a huge old bay fell down by the garage, I spent days at the creekbed sawing it up, stopping often to gaze upward from the dense shadows in the deep hollow. Born a flatlander, I was now a dweller of slopes, every step on our property either an ascent or descent, every glance framed in "up" and "down"—behind us, the hillside rising to its crest; in front of us, the wide-open view of the bay; beside us, steep banks plummeting to the narrow creekbed.

I remember the moment when this new awareness of gradients and slopes shifted my perception in a larger way. A lover of road maps, I could find my way anywhere if the map was good. But as I drove home one day on the redwood-lined freeway that plows a straight furrow across the Oakland hills, something happened. Just past one plunging hollow, as my car pulled up the next hill, it suddenly occurred to me that the valley through which I was passing had been made by water.

Now, I'd been a hiker for years, had crossed creeks and rivers and hills on foot, had dipped far into the Grand Canyon, admiring its water- and ice-carved walls of stone. Of course I knew that rivers whittle away the landscape, that water always flows downhill.

Yet I didn't really know it in my body, did not orient myself by the movements of water. And so I never got it at the visceral level that water creates the landscape. In my mental map valleys were valleys, creeks were creeks, and the tiniest of conceptual fences separated them from each other. It had never consciously occurred to me that at the bottom of every gully flows a stream.

Yet that day on Highway 13, between the Park and Skyline exits, I realized that the valley I was passing through had been carved by flowing water. In fact, every valley snaking downward from the ridge at my left to the bay on my right had been formed by water. This highway was traversing, one after another, the trails that water had created in its journey downward from the ridge toward the Pacific. Water was the landscape architect of everything in sight.

The realization, simple and obvious, dazzled me. In a moment my orientation shifted—from locating myself among these hills by way of roads and streets to finding my way in terms of flowing water. My mental map was redrawn, and the most prominent features on it now were creeks.

My home rested not beside a certain street but on the slope of a certain hillside carved by a particular flow of water. That waterflow was a creek.

My creek.

Peralta Creek.

I had begun to think in terms of watersheds.

⌒

SOON AFTER, I crossed paths with a young elementary school teacher, a bright-eyed fellow with curly hair pulled back into a ponytail. When I told him my watershed awakening moment, his face lit up. "We try to give students experiences like that," he said. In his inner-city classroom they offered watershed education, helping children to notice the flows of water over land. Sometimes they took children out to the nearest creek, showing them how to watch birds or count water insects and explaining how dangerous the paints and pesticides and motor oils are when they run from city streets into creeks to poison the aquatic life.

But one lesson was simpler than all of that. "Do you know about making a watershed with paper?" he asked. I didn't, so he explained. "Take a sheet of stiff paper and crumple it into a ball." Later, at home, I followed his instructions.

After crumpling the paper into a ball, I slowly, gently, teased it toward smooth again—not all the way so that it lay flat, just far enough to set the corners down. I placed the crinkly paper on a lipped cookie sheet.

Then I sat and looked at the wrinkled paper as if it were a miniature landscape and I a pilot flying over it. Hills and valleys rippled across its crinkled surface. Where would I most like to live on this land—on the highest point? In a deep hollow? I took a colored marking pen and drew my tiny dream house—on a hill with a view, of course, just like the house I was living in. Now, where might a forest grow? I colored in some trees with green. Using other colored pens, I drew a neighborhood, then a town, then penned in roads with a black marker.

Now it was time to make rain. Taking a spray bottle filled with water, I misted the paper, watching the rain fall gently on this land. Soon the water began finding its way off the ridges of the crumpled paper and into the valleys, flowing toward the edges of the paper and onto the tray. I had just made a watershed—actually, three of them—and my creeks were now headed toward the sea. Like all waterways, these streams carried pieces of the land over which they flowed. I noticed traces of black in the runoff, where the water had picked up ink, like pollution, from the roads and houses. One of my settlements threatened to get washed out in the sudden downpour.

I remembered what the teacher had told me. "The kids begin to get it when they watch the water turning brown or black," he had said. "They begin to see that what they do on their street connects with the water. They see the relationship."

⌒

ONCE I HAD glimpsed the concept of a watershed, my mind ticked over the Oakland neighborhoods I'd lived in previously, the undulations of their streets. Water's paths had determined the shape of all the earth on which I'd walked. The wide thoroughfares of Grand Avenue and Lakeshore Boulevard, filled with the hum of traffic and the reassuring scurry of pedestrians—wasn't each of them tucked between low ridges? So didn't that mean that each had once been a creek? I looked up old city maps and found that, yes, a creek flowed in a huge culvert below each of them, down toward what was now Lake Merritt. I even found public works employees who dreamed of day-lighting those creeks, bringing them aboveground again, perhaps as green-ways running down the center of each boulevard.

A mile away, in the next-door watershed of Sausal Creek, I discovered a citizens' group busy planting native species and weeding out invasives. They'd recently passed a long-term plan for daylighting the whole of their creek, even in the most densely built-up neighborhoods. A few in the group had started a watershed-specific native plant nursery, collecting dry and wispy seeds from creekside plants in autumn, building a shade structure on a sunny hillside, then planting and tending the seeds in spring. They hoped to grow thousands of plants for a big creek restoration project that would replant the banks of the creek and tempt native trout to spawn there again.

Their plan was carried out a few years later. One rainy winter hun-dreds of volunteers planted tens of thousands of seedlings—willows and currants and bedstraw and wild roses and nettles, all the plants historically found along Sausal Creek and all propagated by volunteers at the watershed nursery. And the trout did return, and their numbers increased year by year.

As the native-plant nursery grew, I put in volunteer hours with the seedlings, moving, sorting, transplanting. I loved the work. It soothed me, allowed my hands to play in dirt.

Delving into creek work, I learned that many people had been doing it for decades already. At home one Sunday, after spraining my ankle near the bottom of our huge stone steps, I spent the day with my foot elevated reading Ann Riley's *Restoring Streams in Cities*, learning about bank stabilization and streamflow and creek hydrology and the curves, or meanders, that a river needs to be healthy. I gulped the book, I was that thirsty.

WE'D BEEN IN the house two years already, and still the old junk was lodged at the bottom of the creek, the tires embedded in clay. I saw the garbage every day when Sapphire and I climbed the thirty steps to the tippy-top of our property and turned left on the one-lane, curving street to follow Peralta Creek gently uphill to its source. That half mile of the creek was stunningly beautiful, at least by urban standards—a sanctuary of deeply shaded bay forest with almost no houses in sight, the urban hum fenced off by tall canyon walls. It was a place to listen to birdsong, where the only other people we met were neighbors walking their dogs or bikers in spandex or an occasional lost-looking driver glancing nervously at the steep drop-off.

And every day on our walk I gazed far down the narrow canyon and spied the junk—rusted iron parts from who knows what piece of machinery, the white and rounded corner of what had to be an old refrigerator, an enormous industrial tire half hidden by ferns and poison oak. Every few months it seemed another load of trash showed up, a mattress or box spring or pile of broken drywall blooming magically overnight.

How could we get that junk out? A few phone calls told me that the city, through its Watershed Improvement Program, would provide garbage bags and gloves—and even send a truck out the following week to pick up the trash, provided that neighbors hauled it out of the creekbed and set it neatly beside the street.

Haul appliances up a forty-five-degree slope? How?

Together, a half-dozen neighbors hatched a cleanup plan. We put out flyers and assigned jobs: one person to pick up the city's supplies, a second to sign in volunteers, a third to make a pizza run for lunch, another to take a group photo, and one more to paint a sign for the photo. Then all we had to do was wait for that Saturday in September. How many would show up to clean the creek?

That week I grew nervous. Maybe we'd get a whole raft of people— thirty or so. No, I'd tell myself, can't hope for that many. In an urban area people don't show up just because someone posts a flyer. Most of them don't even know their neighbors. And even if thirty did come by, how in the world would we coordinate them? The opposite was just as nerve-racking: what if only three or four showed up? Curious, sometimes tense with anticipation, I waited for Saturday.

The day dawned cool with a thick morning fog that in a Bay Area autumn clears away by noon. I ate breakfast and hiked down our long steps to the sign-in table, next to our rickety garage.

Nine o'clock approached, and neighbors trickled toward the table—a few couples I'd never met, two teenage girls, the man next door. I signed them in, sending some to work down the hill and others up our long steps to the little street running through the bay-shaded canyon. After the people stopped arriving I joined a brigade to roll tires through the brushy creekbed, one person to the next, toward the pile building beside the street. Now and then a few more people arrived to sign in, and I sent them up the steps to the canyon, where the rest of our committee would put them to work. Pretty good—maybe twenty or so, all told.

We were so busy pulling and pushing that before I knew it noon had arrived. On my way to pick up the pizza, I drove through the other work area, in the upper canyon. Passing under the first of the overhanging bay trees, I could hardly believe my eyes. Already the shaded road was lined with bags of garbage, small appliances, and piles of tires. The ten or so people I'd sent up here couldn't have done all this by themselves. In the distance a woman with a huge video camera on her shoulder was pointing it at one of our neighbors. Apparently we were making the Bay Area news.

A pickup had been parked perilously close to the drop-off with a rope tied to its tow hook. I heard, "One, two, three, *pull!*" A pause, then, "One, two, three, *pull!*" I had to get out of my car to peer over the edge, where I saw five or six people positioned along the length of the rope, inching an old stove upward with each heave. Far below, Anthony was wrestling an engine block from a trash pile in the creekbed, ready to tie it onto the rope.

I looked left and right and gasped: the hillside was teeming with people. Many more than those I'd signed in were scattered along the length of the canyon.

Then it hit me: there must be *fifty* people here! Neighbors were smiling, sharing garbage bags with people they'd just met, dragging tires uphill in pairs. Some were tackling ropes of ivy that for years had been strangling the trees, cutting it down and lugging mounds of it up to the street. People were laughing, calling out to one another, congratulating each other on the heaps of trash accumulating at the top. We all—neighborhood politicos and introverts, the vocal ones and the those who would disappear quietly at day's end—were working together on behalf of this shared treasure, this piece of Earth that none of us owned and all of us enjoyed.

By the end of the day, a hundred tires were heaped by the road, along with enough rusty appliances to outfit a small apartment, some auto seats, a motorcycle frame, and too many cans and bottles to count. And the industrial-size tire—four feet wide and two hundred pounds, not counting the muck packed inside—got special treatment. Late in the afternoon a large backhoe rumbled slowly into the canyon from who knows where, attached a tow line, pulled out the tire, and rumbled away.

After everyone had gone home to rest their backs and shower off the poison oak, I walked the length of the canyon surveying the mounds of trash. I felt full with the jubilation that had flowed among the neighbors, was lifted almost physically with the memory of laughter and goodwill reverberating up and down the wooded lane.

This was the most significant thing I had done, or could ever do, with my life.

～

ECOLOGISTS USED TO speculate that we'd find the limits of the natural world either through taking or through giving—extracting too many natural resources or exuding too much waste. We don't have to speculate anymore. We've found the limits of both. Our habits of taking and giving are in a race to kill us.

At the moment, waste is winning. Choking a creekbed with trash is but one tiny way of not taking responsibility for what we give out. It is one way we show we have forgotten the breathing world.

In a self-sustaining system, for every breath drawn in, there is a breath released, and the breath is shared among all. What one species produces, or emits, must be able to be used by others. Waste from one is fuel for another. This is how we depend on one another.

The trouble is, when one set of humans decided that the Earth was mechanical instead of alive, they stopped closing the circle of breath. They

breathed in and in and in and said to hell with the out-breath, which basi-
cally meant to hell with everybody else—all the species who depend on human wastes and by-products to be healthy for them. Disregarding the need
for balance between themselves and others, they figured out how to make
things, like plastic, that no one else could use. *Nonbiodegradable* means
immortal. And from the point of view of an organic Earth, *immortal* means
worthless—more like dying than living, and far less useful. Each thing that
provides no food for others subtracts some energy, some fuel, from the
community as a whole.

I'll never forget Michael Pollan urging an audience not to shop in the
middle of the supermarket. "The stuff in the middle is immortal!" he said
with a laugh. He'd kept a Twinkie in his office for two years, and still it
showed no sign of decomposing. "What causes rot? Bacteria and fungi.
They're not interested in foods with little nutrition in them. They know
something we don't." Stay away from the immortals, he said. Food isn't
food unless it will rot.

What we have done in unearthing fossil fuels and turning them into
plastics and preservatives is to produce things that nourish no one. Even
worse, the bottle caps and balloon fragments and bits of carpet are mistaken for food by birds and other creatures, who ingest them and then
starve to death on full bellies.

Through fossil fuels we subtract energy from the system in another
way as well. In hydraulic fracturing (fracking), billions of gallons of freshwater every year are mixed with toxic chemicals and pumped a mile or
more underground under pressure to break rock and release natural gas.
Recovered liquid is so polluted it must be disposed of as toxic waste, so the
water used in fracking is gone from use forever. In addition, groundwater
may be polluted by the toxic seepage. This is a grievous theft from the
freshwater cycle and from all of us—humans, plants, animals, soil microorganisms—who depend on increasingly scarce freshwater for life.

But our other waste from fossil fuels is having an immediate and catastrophic impact: we are breathing out more carbon dioxide than others on Earth can breathe in. Unless we find a way to reverse this calamitous breath, to curb our carbon dioxide emissions and turn them into fuel for ourselves or someone else, the effects on Earth's climate of our heedless breathing out will last for thousands of years, long after humans likely can no longer survive here.

Though I didn't make the connection at the time of the creek cleanup, tackling a problem of waste is indeed the most important thing human beings can do at this time. "Maintaining a balanced and pure human breath also ensures the purity and health of the breath of the natural world," writes the Rarámuri anthropologist Enrique Salmón.

Breathing in and breathing out, the simplest form of meditation.

A creek cleanup as cross-species meditation, a way of tending to the human out-breath, of respecting the in-breathing needs of water and soil, microorganisms and magpies.

⌒

MY CLEANUP HIGH lasted for days. I'd simply never felt such a stretch of satisfied joy.

The group of interested neighbors became, over months, a group ready to preserve the stretch of green along the creek. When a parcel of land close to my house came on the market and surveyor's stakes were pounded in, the time seemed right to take the next step. This was an uphill lot with a particularly spectacular show of native wildflowers at pedestrian eye level each spring. A few adjoining neighbors decided to raise the money to purchase the lot ourselves. An artist friend created a huge poster of a fundraising thermometer; I built a wooden backing and mounted the poster; and two or three neighbors gathered to dig the post hole, raise the heavy sign, and tether it with guy wires. Then we stood back to admire. The poster announced that we intended to save this land. That we valued its company for the wildflowers and oak trees it grew. That we wanted everyone in the neighborhood and community to be able to enjoy it this way forever.

As the fund-raising thermometer rose, the small group of interested neighbors grew: a neighbor who'd been working for years already to pre-serve the canyon; a cyclist from another part of town who had fallen in love with the canyon while biking through it. Everyone knew that this tiny lot was just the beginning; the whole canyon needed the same protec-tion. Decades before, the canyon had been parceled out into more than a dozen lots with as many different owners, and in recent years these lots were drawing a lot of attention from developers scooping up even the most steeply sloped properties. Engineering techniques had evolved, and inves-tors thriving on real estate and dot-com booms were easily coughing up the hundreds of thousands of dollars needed to erect massive retaining walls. The neighbors needed a more powerful tool for preserving the canyon. We decided to become a nonprofit land trust.

In June of 2001 I filled out the simple form that brought the new entity into being and mailed the required thirty dollars to the state. The call that began with a simple voice in my ear announcing "Your house is here" led to the forming of a neighborhood land trust devoted to buying the remain-ing undeveloped parcels in the canyon and preserving them in perpetuity.

At the organizational meeting for the board of the new nonprofit, I explained the challenges we faced, the interminable process of getting tax-exempt status, the endless operating expenses we would have to raise. The rest of the board listened politely then, one by one, said, "We don't know exactly how to do it, but we want to purchase the land and keep it green." They spoke of the construction happening just down our street, where the side of a tall hill had been dug out and trucked away, an enormous divot of a wound making room for a foundation and impossibly high retaining wall. My next-door neighbor expressed the feeling of the group when he said, "You know, in this city the trees are protected. You have to get a permit to cut one down. But no one protects the ground."

~

I WAS THE founder and first president of the land trust, and I led the board in those early months as if I knew what I was doing. In fact, I had little experience working in organizations. I'd been a self-employed student for almost twenty years, had never done battle in office politics, and only rarely organized large events. I didn't have a clue how to build a group from scratch.

Yet I seemed to know how to take each next step. I began networking with land trust officers in the wider Bay Area, a band of crusty and congenial professionals whose combined years of experience in land scuffles added up to centuries. I reached out to the Friends of Sausal Creek in the neighboring watershed, approached possible donors, applied for grants from local foundations. I spent hours on the phone chatting up biologists and biodiversity specialists, the national land trust network, neighbors concerned about overdevelopment. My contacts list grew along with my knowledge of riparian corridors, native plants, and migratory birds; real estate contracts, charitable foundations, and grant writing; budgets and strategic plans. I navigated the world of urban creek preservation as if I'd been born to it.

After six months of operations, the board had a solid vision and a large pile of paperwork to show for its efforts. Yet neighbors remained skeptical: Buy up all those properties? "Where," they asked us, "will you ever get all that money?" By *you* they meant "you hopelessly romantic little scrap heap of a group." Or at least that's what we heard. We were all too aware that we could barely cover even our meager operating expenses. As the weeks crawled by, each piece of good news—a new donor, a write-up in a local paper, a five-thousand-dollar grant, an interview on KCBS radio with its million listeners—was met by its opposite: another skeptical neighbor, another grant application turned down, another developer offering hundreds of thousands for yet another lot.

Daily during those months I walked with Sapphire through the canyon. Its arching bay trees, its pair of tall and stately redwoods, its hillside of currants and yarrow were now friends. I asked the bay trees and the dry creekbed—it was summer now, the dry season—how they wanted us to proceed. Each time I hiked the half-mile length of the canyon I spotted a native plant I hadn't seen before, a new wrinkle of light dappling the green understory. From each day's walk I also carried away a renewed conviction: this canyon wanted to be preserved. The creek had a will of its own, and its will was to continue to run free, aboveground. The canyon intended to remain a semiwild haven for owls and mice, songbirds and raccoons, and the humans who walked or biked along its shaded, quiet length. And it would use humans to exert its will.

Each time I contacted this sense of purpose, I was dumbfounded. Not only did the canyon have a will, it seemed able to accomplish its will in the world. The canyon was an actor, recruiting humans or others to its purpose, actively preserving its own life. Each time I inquired of the canyon, I felt supported by more-than-human forces.

I had never contacted the will of a place before. When the eagle had flown out of its way to greet me, I may have been forced to recognize a cocreator, but I'd hardly allowed that line of thinking to change my life. I still regarded most nonhuman beings as recipients of human action. Animals moved according to genetic codes and physical laws, their acts limited in scope to their piece of forest or prairie. Grass and oaks and rocks were at most watchers, witnessing the wisdom or, more likely, the witlessness of human acts.

Yet now I flirted with a new possibility. Perhaps it was because the stakes were so high—the potential destruction of a place that I and many other people enjoyed and found solace in. Perhaps it was the odds against reaching our goal—a small neighborhood raising millions to purchase more than a dozen outrageously priced parcels. Perhaps it was because I needed to find a new way of being effective, since twenty years of working only with words had done little to build my confidence that I could get things done in the "real" world. Or was I, after years of meditative work, contacting one of the mysteries of this Earth? For whatever reason, I allowed myself to entertain the thought that the canyon was acting on its own behalf (and ours). And so, despite all discouraging evidence, I remained calm—much of the time—about the canyon's future. Our small land trust was but one partner in this effort to preserve a semiwild green space; the other partner was the place itself.

Of course, sometimes discouragement tackled me too: we didn't have the more than a hundred thousand dollars needed for this first—tiniest—lot, let alone the millions it would take to purchase all the canyon parcels.

One night during those early months I had a dream. In the dream I was taking part in a ritual circle—a few people setting treasured objects in the center of the circle and talking about what they symbolized. At the end of the ritual each participant could choose an object to take home. I headed straight for the one I knew was mine: a replica of the two-inch-high Day of the Dead skeleton I'd purchased recently (in waking life) in New Mexico. The dream skeleton was a quick-moving, foot-high fellow with a sombrero on his white skull and a red and blue serape draped over his bony shoulder. I approached him before he could scurry away, crouched down on the floor next to him, and poured out to him the issue of the canyon, weighing heavy on my heart. He listened, then responded slowly in a hoarse, gravelly voice: "You've got to give it time. These things have their own timing."

~

In 1995 THE highly respected American environmental historian Bill
Cronon wrote a piece called "The Trouble with Wilderness" in which
he criticized not wilderness itself but the ideas that tag along behind it.
The idea of wilderness, he said, preserves "a dangerous dualism" between
human beings and nature. The dichotomy has long been there, from the
Greeks who valued only the nature they had tamed into gardens to the
American settlers with their scorn and contempt for the "howling wilder-
ness" of forests and untilled land.

But in the latter half of the 1800s a dramatic about-face took place.
After settlers had damaged each successive bioregion along the frontier,
Americans turned immediately to longing for what they had just destroyed.
The dualism between humans and nature remained but was now reversed,
and the new ideal became pieces of "untrammeled" ground where "civili-
zation" could not encroach.

The idea of untrammeled ground was, of course, romantic. And false.
Americans had to force Indians out of places like the Merced River Valley in
California to set up national parks like Yosemite. The whole continent had
been walked on and much of it modified by Native people. However, the
changes Indians had wrought in the landscape were subtle, and European
Americans often could not see them.

The idea of wilderness is seductive, Cronon suggested, a siren song
tempting us to escape from "tainted" cities to lands imagined as pristine
and pure. We go someplace else to romp or ponder, as if wilderness were
our real home, instead of taking responsibility for the lands where we live.
In effect, we write ourselves out of nature. It exists only where we aren't.

Cronon was right: a streak of romanticism flows through the environmental movement, an idealizing of wildness—as long as it is elsewhere—
that prolongs our stubborn sense of separateness from nature.

But something about his essay has always bothered me. Something is
left out, something big and silent, the elephant in the room.

That something is capitalism, the driving force behind the English legal definitions of landowners' rights. In Cronon's essay capitalism receives only incidental mention, and private property none at all. Yet an idea of wilderness was needed in the first place because landownership is defined, à la Hobbes and the social contract, in terms of an individual owner accruing profit.

In English legal tradition, owning land bestows the right to do pretty much anything you want with it if by doing so you can accrue the greatest profit. A property title is a legal permit to destroy the land if you so choose. (Or at least it was before the era of environmental laws.) The "highest and best use" in real estate law is defined not in terms of the land's benefits but in terms of property value—a system skewed toward economic gain, especially the kind that involves altering or defacing the landscape or destroying its ecological integrity. "Improving" the land means making changes that increase property value, such as building structures or roads or power lines or factories or mines—all the things that damage the surface or destroy the interior, rendering the land unable to function as a healthy ecosystem. It's a tradition that pits the rights of individual landowners against the rights of others who also have an interest in the land, such as human and animal neighbors, soil microorganisms, those who breathe the oxygen produced by the vegetation, and of course all the downstream users of water. It's also a system that gives precedence to the owner's economic right to profit but traditionally ignored the economic value—to either the owner or the neighbors—of clean air, water, and soil as well as the aesthetic benefits of landscape, let alone the inestimable value of a functioning bioregion.

English-language land law tended to define owners' rights in all-or-nothing terms—absolute rights if you own title, none at all if you don't. And naturally such a system bred its opposite, a wilderness law prohibiting any kind of land use. Cronon was right: the idea of wilderness prolongs the all-or-nothing approach. What he failed to mention, perhaps because he thought it so obvious, was that the idea of wilderness arose in reaction to centuries of land destruction based on private profit. Unfortunately, the idea of wilderness merely reversed the old rule, playing the same all-or-nothing game.

We need a new, in-between ethic, Cronon suggested, a way to recognize wildness closer to home. We need to value the tree in our backyard, not merely the one in the national park.

In the land trust movement I discovered one such in-between ethic. It was a method that worked within conventional understandings of private property, accepting the notion that the rights of the landowner trump those of all others. (And so I remained restless with the strategy even while using it.) But land trusts also subverted the old ideas about property by emphasizing the economic value of land features that were traditionally undervalued, such as ecological integrity or aesthetic beauty, and so they brought into the equation the rights and interests of neighbors and the wider community.

In 2001 when our small group came into being, the land trust movement was already huge and growing fast. We were the newest of about 1,200 such groups nationwide. But we were one of only a handful doing so in the middle of an urban area—not outside its boundaries but within city limits. We were trying to practice what many cities worldwide have long known: that plots of wildness scattered throughout neighborhoods increase the quality of urban life. In an ideal city, every resident has walking access to some place, however small, where nature gets to do its thing in gardens or parks or along rivers or creeks, where the urban roar is muted by trees and vegetation, where human "improvements" do not dominate the field of vision. In our tiny shaded Oakland canyon, people could enjoy a few moments of solitude, which meant not being alone but trading the ubiquity of human noise for the company of bay trees and titmice and trickling water. Under the canyon's sheltering canopy of bay trees, we walked and breathed and listened and enjoyed.

The Kiowa-Cherokee writer N. Scott Momaday wrote, "You say that I *use* the land, and I reply, yes, it is true; but it is not the first truth. The first truth is that I *love* the land; I see that it is beautiful; I delight in it; I am alive in it."

In trying to preserve our little stretch of creek, we were putting enjoyment at the top of the list of land values. We were voting for love and delight.

⌐

SIX MONTHS AFTER the land trust filed its articles of incorporation, we witnessed a miracle. A family who owned three creekside lots in the canyon decided to donate two of their parcels to our fledgling group. They would trade their right to "improve" the land for some federal tax breaks.

Instantly the land trust was on the map—the real estate map—and our assets, at Bay Area land values, totaled a quarter of a million dollars. At the dedication ceremony in the canyon shortly after, friends and neighbors stood on the narrow winding road over the creek while a doe browsed her way carefully through the canyon below us, quietly watching.

But progress remained slow. It seemed that the land trust could never make headway on those for-sale canyon lots. And every passing year raised the price of those lots by sometimes hundreds of thousands of dollars. Yet new people kept finding us: a young, impassioned community activist; a real estate attorney; a professional fund-raiser and grant writer. Another parcel was donated, but as the real estate bubble expanded, more cantankerous neighbors threatened to sell. Surveyors' stakes appeared regularly in the canyon, along with new FOR SALE signs. Yet one sale after another fell through. The months crept by, and the canyon remained a green haven.

~

I'VE BEEN GONE from that neighborhood for the better part of a decade, and in that time the land trust has been busy. In 2010 it completed its original mission and put into conservation the last of the thirteen unprotected lots. But the nine years of work were full of ups and downs. "It was like Mr. Toad's wild ride," said Dave Barron, the real estate attorney who led the group for the last seven of them. "Every time a lot came up for sale, we'd send out this panicked broadcast to the neighborhood, asking for money."

The final price tag for preserving all those lots? Nearly a million dollars—some of it raised from donations and used to purchase lots held in the land trust's name, some received through city bond funds set aside for urban creeks, and some raised to purchase conservation easements for protecting creekside stretches from development. The group has scrabbled, bargained, and negotiated for that million dollars and along the way has won tens of thousands of dollars in grant monies. To this day neighbors continue to hold annual canyon cleanups and creek restoration workdays.

I'm not the only one who has noticed the canyon's strong will. "In some respects," Dave Barron told a Bay Area TV crew in 2010, "the canyon seems to protect itself—in the way things have happened, some of the breaks we've had, some of the luck."

As I watch his interview, I chuckle to myself. So Peralta Creek is still having its way.

9

Shvana

The mountains, I become part of it . . .
The herbs, the fir tree, I become part of it.
The morning mists, the clouds, the gathering waters,
I become part of it.
The wilderness, the dew drops, the pollen . . .
I become part of it.

—NAVAJO CHANT

I HEARD HER THREE legs clumping slowly down the hall. *Ka-thump-thump*. Pause. *Ka-thump-thump*. Pause.

That was odd. Lee was in the kitchen at the other end of the house. Why would Shvana be leaving his side to come toward the guest room? Shvana loved Lee. Passionately. It was the electric devotion of a brand-new parent—heart-stoppingly tender and full of awe—except Shvana could carry it off 24/7. She had sustained it already for many years.

Shvana was a peppy short-haired black terrier with a tan face who had been with Lee, a yoga teacher, long before my friend Jackie met and married him. Shvana—her name means "dog" in Sanskrit—was smart, good-natured, and sweet, and she knew her purpose in life. Around Lee she carried herself with the confidence of a royal chambermaid: the king might sleep with others, but only she knew where he kept his underwear and exactly how he put his trousers on each morning.

Shvana knew, in other words, that Lee's welfare depended entirely on her. And so when she developed bone cancer in a foreleg and Lee and Jackie, agonizing over whether to put her through the terrible treatment of amputation or to put her to sleep, meditated with Shvana to ask her opinion, they got the feeling she wanted the treatment: take the leg if it would allow her to be with Lee a little while longer.

Which is why I was surprised to hear her heading toward the other end of the house. Sapphire and I were in the guest room, freshly arrived in L.A. after a six-hour drive through the Central Valley—all blinding sun and dry heat—and I was looking forward to resting my eyes in the cool, dark room. Lying down, I found myself wide awake instead of sleepy, and in the siesta-quiet house I slipped more easily than usual into a relaxed awareness.

Maybe now was the time to consult with Shvana. This trip had been planned, after all, with Shvana in mind.

᠀

IT WAS 2003, and a lot of changes had taken place at home in Oakland. The land trust was now a couple of years old, and to tell the truth, I was feeling restless at the helm. The organization had survived its infancy, and the steep learning curve—the fun part, to me—was over. A slow and steady attention was needed now, and I wasn't nearly as fond of this stage of things. And not very good at it either.

In the past year my personal life had undergone upheaval. Six years earlier Anthony and I had started out full of hope, but we'd run up against differences that grated rather than polished us. Eventually we were scraping open old family hurts instead of healing them. "You're stuck in the past," Chuck the medicine man told me at every ceremony. "Way in the past. We're helping you get out of the past." He meant it literally, and I should have been taking him that way. Every argument with Anthony carried me back decades, shrinking my sense of self down to child size. It had taken me a long time to come to grips with the truth, that leaving the past in this case meant leaving a troubled relationship. Finally, a year ago, Anthony and I had ended, and I'd fled to L.A. for a few days with Jackie and Lee.

It had been a grueling drive a year ago, after not sleeping a wink the night before. Pausing at a lonely rest stop along Interstate 5, I meandered with Sapphire across windblown dirt and heard commotion in a scraggly tree. Looking closely, I discovered the branches teeming with orioles on their spring migration. I thought back to the baby oriole of—what was it already?—a decade before. Ten years almost to the day since I'd found him fallen from his nest. I'd hardly seen a single oriole in all the years since. Yet here they were, showing up in a flock as I fledged alone again into the world.

Now I too lived among a flock. After Anthony moved out, roommates moved in, and my once-quiet treehouse now hummed with activity—a trio of thirty-year-old young men coming and going, their assorted friends and girlfriends dropping by. Sapphire and I might come home to beer and burgers on the patio or the sounds of conversation down in the kitchen. "I've become a den mother," I told my friends. I was forty-six. One roommate, a rough-and-ready anthropologist, loved Sapphire so much he couldn't pass her by without stopping each time to stoop low and smooch the top of her furry head.

Though I was enjoying life with roommates more than expected, I needed a retreat now and then with Jackie and Lee. And their dogs. Sapphire and I would pile into the car, I'd put in a book on tape, and six hours later we'd play with our friends in L.A. Sapphire loved the trips; they gave her time off from looking after everybody and she could relax. I loved them for the same reason.

⌒

A FEW DAYS before, Jackie had called me on the phone. "Could you talk to Shvana while you're here?" she'd asked. Shvana was slowing down. Months before, when her leg was first amputated, she'd gotten around remarkably well. But now her pointed black nose hung a little closer to the ground, and hopping across a room seemed to take a great deal of concentration. Not that she had to hop much; Lee usually picked her up and carried her out the patio door to do her business on the lawn. But she was determined to follow him everywhere, and clearly this was getting harder.

Recently, instead of waiting to be escorted out the door, Shvana sometimes peed and pooped on the carpet, staring directly up at her people. Was she trying to send a message? Or was she just incontinent? Was it time to put her to sleep? With a pet, how do you ever know? Lee and Jackie had reached that wrenching stage, wanting to prevent her from suffering, not sure it was her time to go, and definitely not yet ready for the end. Would I, an honorary auntie, talk to her, see what I could find out?

Of course I would. I didn't know Shvana very well, since during our visits she stayed near Lee and paid little attention to me. Yet checking in with a dog or cat was comfortable for me now, with Sapphire honing my abilities daily. From time to time I would catch Sapphire following my thoughts—even, I swear, planting one. Like the time I was writing in my journal, idly gazing out the window, and suddenly found myself thinking of dog biscuits. Turning back toward the room, I found Sapphire sitting at attention and staring directly at me. *Okay, okay.* I got up and gave her a biscuit.

As in any relationship, the quality of attention seemed key. Trying to talk with an animal—or human friend, for that matter—while distracted by work or emotions or thoughts or fears about the future or any of the countless mental disturbances that Buddhists call "monkey mind" always put the kibosh on a genuine connection. Conversations, whether verbal, with humans, or in pictures, with animals, went most smoothly when the mind was open like the sky, able to notice the passing clouds yet not get caught up in them.

Not that I was good at any of this. My mind tended toward the turbulent end of things. Especially these days, feeling the stress of changes I couldn't yet name. My mental sky hosted lots of massive gray clouds, moving fast. I latched on to every one of them.

Still, I practiced breathing and letting them pass, trying to reach that place that Shunryu Suzuki said is "nothing special." Before you go to see a famous mountain range or beach, you think it will be thrilling. "But if you go there you will just see water and mountains. Nothing special." Enlightenment, the mind like an open sky: before attaining it, people think it must be magical. Yet after, they find it is nothing.

"But yet it is not nothing," Suzuki added. "For the person who has it, it is nothing, and it is something. . . . Do you understand?"

No, I didn't, not really. All I knew was that meditating calmed me a bit, helped me see more clearly, nudged me toward a state of mind where I could appreciate close friends and satisfying work and a regular yoga practice and all the other good things I was enjoying even as the ground shifted under me. Meditating also kept the mind a bit freer of projections, more able to hear, in conversation with another, what the other was truly saying instead of what I wanted or expected to hear.

So on this particular afternoon, when I found myself calm and alert after the long drive instead of sleepy and distracted, I lay down on my back to meditate and decided to say hello to Shvana.

She was at the other end of the house—with Lee, of course. So I brought an image of Shvana to mind and turned my full and open attention toward it.

Instantly something happened. In my mind I heard words: *I don't plan it*. There was a feeling of urgency. *Tell them I don't plan it. I don't plan it. I don't plan it*. The thought just kept repeating, the intensity rising. *Tell them I DON'T PLAN IT!* Not angry, just the most insistent voice I have ever heard. *I don't plan it. I DON'T PLAN IT!*

Okay, okay, I got that, I tried sending in response with a feeling of reassurance. I could only assume, since Jackie asked me to find out why Shvana was soiling the carpet, that Shvana was referring to peeing and pooping. Her way of saying it was all accidents.

Still the words repeated: *I don't plan it. I DON'T PLAN IT!*

And that was when I heard Shvana lumbering down the hall on her three aged and awkward legs. To leave Lee, I knew, she had to be headed somewhere truly important. Then she entered my room. It made sense because this was also Lee's study, and Shvana probably intended to lie down in her favorite spot.

But no. She hopped slowly, painstakingly, to the side of the guest bed and stopped. I heard her pause there, but more than that, I *felt* her. Lying on my back with my eyes closed, I was aware of her in the way a flower must be aware of the sun, the command of the light so inexorable that the flower can do nothing but pivot on its stem to face it.

So I opened my eyes and sat up. There she was, staring at me hard. I leaned over to pet her.

Again the stuck record: *Tell them I don't plan it. I don't plan it. Tell them I don't plan it*.

I tried to reassure her. *Okay, Shvana, I hear it loud and clear. You don't plan it. I'll tell them you don't plan it. I got it, Shvana. You don't plan it*.

The urgency subsided, but still she sat there looking at me, as if waiting for more.

So I asked, *Is there anything else you need? What would make you more comfortable?*

I got the feeling she wanted her spine massaged. Of course—it was probably sore from her off-kilter hopping. I rubbed gently up and down both sides of her spine, and she settled with a sigh into the floor. But she kept her eyes open, looking my way.

After a while I got tired of leaning over the bed, and I lay back down to rest. But Shvana did not budge. She had come all this way down the hall to talk and, dammit, she would *talk*.

I've rarely felt such a wide-open invitation—from humans let alone dogs—so I decided to press on. By now I realized I didn't have to work hard to make my questions into pictures. Shvana was skilled at this way of communicating. In her presence there were no distractions. Her mind-sky was clear. She knew what she wanted, and she would make others around her know it too.

So instead of translating into simple childlike language or images, I sent thoughts in my mind as if I were speaking them to another adult. *Lee and Jackie love you very much. They want to know if it's too hard for you, getting around. Are you ready to leave your body? Do you want to go on?*

I had no idea if such words made sense to her. Or even to me. Was there such a thing as "going on" or "leaving the body"? Many people talk that way, but who really knows what happens, if anything, after we die?

Even as my mind wandered I heard a prompt reply, in the tones of a seven-year-old who has made a firm decision. *Oh, I would never choose THAT.* Pause. *But if Lee chooses it for me, then it's fine.*

No questions. Matter-of-fact.

Okay, Shvana, I'll let him know.

Tell him I don't plan it. I don't plan it! She'd returned to peeing and pooping.

Okay, Shvana, I'll give him the message. I sent her a picture—more a feeling—that I was pressing the message into him. *I'll make sure he understands.*

The sense of urgency calmed again, and it seemed like our talk might be over. I prepared to settle back into my own thoughts and get some rest. But Shvana's presence still loomed large. Now that she had me in the terrier jaws of her attention, she was not about to let go.

Okay, I thought, she really wants to be doing this, so why not keep going—ask her something I've never had the chance to ask anyone, human or otherwise?

Shvana, what do you think it's like to die? Could she understand such an abstraction? Was she even listening? *Is it like?*—and I sent a picture of dashing madly on four healthy legs across a green meadow under bright sun, diving into the grass and rolling with abandon.

Instantly, I got a feeling-picture back. *It's more like*—a feeling of melting away, spreading out quietly into everything, like fog creeping across the ground and filling all the air. In that melting-together was great peace, a feeling of deep quiet and intense calm, of all being right just as it is.

I became very still inside, awestruck. Shvana had a clear idea about what would happen after death. Even more, she was ready for it. Peaceful. Unconcerned.

Our talk was drawing to a close. I sent her a feeling of great thanks. Full of wonder, I massaged her some more then lay down and turned my attention again to my own thoughts. A minute later I heard her get up and stumble purposefully out the door to find Lee.

⁓

THIS CONVERSATION WITH Shvana was magical, though I didn't feel it at the time. Like many exchanges with friends, this one seemed ordinary, matter-of-fact, and I didn't realize until later that it would remain in my memory as one of the clearest and most profound conversations I would ever have with anyone, human or other. It was simply a straightforward exchange with someone I didn't know very well. A quiet conversation. Meditative.

Of course, there's plenty of magic in meditation—and especially plenty to be found *after* clearing the mind—but the real magic is that of opening the heart. Becoming receptive enough to hear another. Entering with empathy into another's experience.

Listening, the other *L*-word.

That evening I told Jackie and Lee over dinner what I had learned from listening to Shvana. "She doesn't plan it. She needs you to know she is just incontinent." They were quiet, nodding.

Almost apologetically I told them the rest of our conversation, still not sure I wasn't just making it up. "She says she would never choose to leave Lee, but if Lee chooses it for her, then it's fine." Such single-minded devotion! My voice softened. "And after dying, she thinks it will be like melting into the whole." I felt tears stinging my eyes, and across the table Jackie's eyes too were filling.

It is always like this. Listening deeply brings a feeling of softness, the doors of the heart sliding wide open in the chest.

⁓

WE LOVE OUR pets for their ability to bring us back to the present, to remind us to laugh and roll in the grass or luxuriate in a sunny spot on the floor. Animals communicate using so much less language than we do, which might well be a source of their joy. "The spotted hawk swoops by and accuses me," wrote Whitman; "he complains of my gab." For good reason. We're a talky bunch, relying on language in almost every waking moment.

Yet words can also get in the way. When we are in love, they may say too little; when we are in conflict, they say too much. Which may be one reason the Dagara people of West Africa are a bit skeptical of language. It is only, they say, a distant echo of the meaning to be found at the center of all-that-is. "*To utter* means to be in exile," writes Dagara native Malidoma Somé. Language is proof we do not reside at the source. "Language implies nostalgia for our true home."

That home—that fullness of meaning—is nature. At the center, writes Malidoma, there is no distance between thought and deed. If we resided at the center, "we could see, feel, and touch the results of someone's thought instead of relying on words to give us a picture of it; thought would instantly produce the thing." Words would not be needed because meaning is ever-present. At the center there can be no sentence, no space between subject and object, for noun and verb reside in each other.

The Dagara believe it is possible to tell how intelligent a being is—how close to the center of nature—by how much language it uses. The beings closest to the source are plants and trees, for they do not depend on language to communicate, and so they are the most intelligent. Farther away from the source are animals, who need some language to survive. And farthest away of all are human beings, who depend almost entirely on words. We are deepest in exile from the source.

Western skeptics will pooh-pooh the idea that animals can talk; the Dagara might well be skeptical but for the opposite reason—that animals have little need to talk because they are wiser than us. Perhaps this is why I have always experienced talking with an animal as both startling and softening—startling because we usually look to animals to deliver us from language, to prod us toward our bodies and senses, and softening because the communicating is usually more direct than words. Feelings and pictures take over—less ambiguous than words, more compelling. Closer, in the Dagara sense, to the source.

Most startling of all is talking with an animal about what happens after death—and most softening as well—for what other topic fills us with so much dread?

~

A LONG, LONG time ago in India there lived a boy named Nachiketa. As a child he'd been unusually devout, and now that he was older, his sincerity had grown a teenage edge. Like most adolescents, he would not put up with hypocrisy. So when his father gave away all his possessions just to look good—"to get spiritual rewards," the Katha Upanishad says—Nachiketa called him on it. The cows his father had sacrificed were decrepit, barren, and dry of milk. They were not true gifts. Only giving away something of value could benefit the soul. So Nachiketa approached his father and asked, "Father, to whom will you give me?"

His father, hearing the rebuke in the boy's question, ignored him. But Nachiketa could not leave it alone. He repeated his question, then repeated it again.

Finally his father had had enough. "Go to hell!" he shouted. Literally, "I'm giving you to Death!"

So Nachiketa traveled to the house of Death. Perhaps he was gored by an ox, or perhaps he suddenly took ill; the text doesn't say. In any case, he arrived at Death's door and waited there for three days and nights. Death, it seems, was out on other business.

When Death returned home, he was mortified. A guest at his doorstep who had not been received properly! Guests were the face of God paying a visit, and failing to worship them by offering water and food and by washing their feet was worse than rude; it was a divine offense. Death had to make amends.

So Death offered to grant the boy three wishes, one for each night that his guest had been neglected.

Right away Nachiketa knew his first wish. Like any adolescent who picks a fight but deep down means well, he begged, "Please don't let Dad be mad at me anymore."

And it was done. "When you return home, your father will love you as before. What is your next wish?"

Nachiketa was self-aware for one so young. He saw that he was afraid. Perhaps he'd been unseated by his father's anger, or perhaps he felt jittery about growing up. "Please," he begged Death, "teach me the way out of fear." In another time or place Nachiketa might have been an altar boy or an herbalist's apprentice, but in ancient India the secrets of the universe were housed in the fire sacrifice, so Nachiketa said, "I know there is no fear in heaven. Please teach me the fire sacrifice, which leads to heaven."

This second wish too was granted. Death promised to teach him that ritual from which the world was born in the beginning and from whose fire it is born anew in the secret places of each heart. Death would teach Nachiketa to join inner and outer worlds into one. This was the way to heaven—to living without sorrow, without fear.

He should have been wary, the Lord of Death. A young boy wanting the secret of a happy life, without fear? Death should have seen the final, impossible, wish coming.

"Please, Lord Death, what happens after we die?"

"Oh, no, not this! Ask me for anything else!" Death protested.

Ask for sons and grandsons who will live
A hundred years. Ask for herds of cattle,
Elephants and horses, gold and vast land,
And ask to live as long as you desire.
Or, if you can think of anything more
Desirable, ask for that, with wealth and
Long life as well. Nachiketa, be the ruler

Of a great kingdom, and I will give you
The utmost capacity to enjoy
The pleasures of life. Ask for beautiful
Women of loveliness rarely seen on earth,
Riding in chariots, skilled in music,
To attend on you. But Nachiketa,
Don't ask me about the secret of death.

But Nachiketa would not let it go. Dogged as a terrier, he pressed on. "How can I enjoy any of this," he asked, "when I know my life will end? Everything you are offering is pointless if we all die.

"I have to know this one thing: is there life after death? Tell me the truth."

⌒

THE QUESTION ASKED by a teenage boy is the same one people have been asking in the four thousand years since, the same question I put to Shvana near the end of her life: What happens after we die?

The answers humans have given vary with time and topography. The Katha Upanishad answers Nachiketa's question with a long teaching about how to reach eternal life: realize the union of inner and outer, the great Self and the small self. Find yourself a teacher and get busy, the text says. This will take a lot of work.

Sharp like a razor's edge, the sages say,
Is the path, difficult to traverse.

It will involve practicing intensive meditation so you can recognize the difference between the passing world of the senses and the lasting world of the Self. But at the same time you must learn to recognize the Self in all creatures, "small as a thumb" in the depths of each heart. You must banish all that is selfish within you and learn to reside in the calm awareness that arises from realizing the unity of all. The rewards, says the text, are beyond compare:

> *When all the knots that strangle the heart*
> *Are loosened, the mortal becomes immortal,*
> *Here in this very life.*

Other cultures of the ancient world found other routes beyond death. Egypt, which received the life-renewing floods of the Nile every spring, gave the world the idea of resurrection—and tried to guarantee it for their royalty with embalming.

People around the ancient Mediterranean tended to regard dying as a journey downward to a shadowy underworld. The Sumerians of Mesopotamia gave us the story of Inanna visiting that land on a harrowing descent in which, step by step, she is stripped of her belongings, then her clothes, until she arrives naked at her destination. Virgil in his *Aeneid*, only decades before the birth of Jesus, wrote of two roads in the underworld, one leading toward a place of the damned, the other toward a sweet and pleasant paradise of sunny fields.

In later centuries Germanic people composed verses about Valhalla, Christians about Heaven with its city paved in gold, and Muslims about the rapturous Garden of sweet fruits and fragrant flowers.

But Shvana's answer to the question of what happens after death was not quite like any of them. She did not seek eternal life in the sense of continued existence as herself. She did not expect to run again as she had when she was young and strong. I had no sense that she secretly wished for her striving or suffering to end. It was not bliss she was after but simply a different state of being. Shvana expected to melt silently into the All, as easily as fog dispersing through air. To her it was quiet and peaceful. Nothing special.

⁓

I THINK OF Mark Bittner, whose story of befriending the wild parrot flock in San Francisco is told in the movie *The Wild Parrots of Telegraph Hill*. Mark took the sick parrot Tupelo into his apartment to nurse her back to health and was cuddling with her one night before going to sleep when he felt a wave of gratitude that was not his own. He experienced it as coming from Tupelo, some thankfulness for the connection between them. Then, as he put her on the floor for the night, another wave of feeling, this time sadness at being parted mixed with resignation. The next morning he discovered her body curled up near the heater; she had died in the night.

Bittner muses on a teaching from Shunryu Suzuki about seeing a waterfall at Yosemite. At the top, Suzuki Roshi said, the river approaches as a whole, as one river. When it tumbles over the edge, the river breaks up into drops, millions of individual drops, all falling separately over the rocks and through the air. At the bottom the drops come together again, no longer individual drops but a river. "That waterfall—that time in between the top and bottom—is our lifetime," Mark says quietly. "And all the individual little droplets think they really are individual little droplets until they hit the bottom, and then they're gone.

"But that droplet doesn't lose anything. It gains. It gains the rest of the river."

⁓

THE ANCESTORS RETURN in the drops of rain that moisten the parched earth, say the Indians of the arid Southwest. It is true at the literal level, of course; we are made of the molecules of all who went before, back to the primeval asteroids. "I become part of it," chant the Navajo, placing a sick person at the center of a large sand painting on the ground—a cosmic map—to restore the person to health. To fall out of harmony with creation is to fall into illness; to recover is to restore balance among all the competing forces.

But the old ones return in spirit too, watering with love and care their descendants now alive. The people are not separate from the land, wrote Paula Gunn Allen. "The Earth is the mind of the people as the people are the mind of the earth." It is not "mere 'affinity' for the Earth. It is not a matter of being 'close to nature.' The relationship is more one of identity, in the mathematical sense." The same as. Completely overlapping sets. Land, people. People, land. Good health requires walking in awareness of that oneness.

In how many ways does my culture resist this identity? I think of our burial system, dependent on millions of gallons per year of chemical poisons. (Though poisons are usually thought of as bringing death, they may resist it too, preventing tissues from undergoing a life-renewing rot.) Embalming in this country, which began during the Civil War as a stopgap measure to allow bodies of soldiers to be shipped to their families, turned into a multimillion-dollar industry geared toward ensuring that flesh does not become a part of earth. Every year we bury a million tons of steel coffins underground, or enough to build another Golden Gate Bridge, and the amount of concrete made into burial vaults each year would pave a two-lane highway from New York to Detroit.

What will it take for us to know our identity with Earth, to love where we are so deeply that we can look forward to becoming part of it at death?

❧

"THE FANTASY OF transcending death is opposed to everything I care about," says the philosopher-scientist Donna Haraway. From biology as well as from feminism, she learned friendlier attitudes toward the body, including an unequivocally positive view of bodily death. "Without mortality," she adds, "we're nothing." If there are no limits, nothing new is needed. Mortality breeds innovation. Limitation presents "a condition of possibility."

Haraway's acceptance of death is one of the things I've always appreciated about her work. It is something I too learned from feminism, as well as from my own illness—some patience with limits, especially those of the body, some compassion toward the weakness that will ultimately end in death. If I have gone kicking and screaming through the deaths of others, I have rarely been frightened of dying myself. From what I can glean from reports of near-death experiences, including those of friends, the process of dying will be peaceful, probably even tinged with bliss, whether or not there is physical suffering. Those who have nearly drowned report the utmost quiet and peace; so do those who suffer strangling or shooting or stopped hearts or mangled limbs. Only a very small percentage of people report feeling frightened while dying.

But there is another aspect of dying that terrifies me. Losing my self, the annihilation of everything I consider to be "me"—that one unhinges me. When Death skulks nearby, touching someone I know with serious illness, I can wake terrified in the middle of the night, heart pounding to the point of panic. Yes, I'm frightened of losing that person, but even more, I'm terrified of losing me. Someday I will be disappeared. Gone.

I'd be happy to go through the transition of dying if I could retain my identity on the other side. (Note the euphemism of "transition," the stance of bargaining. I don't really have to *die*, do I? Something of me will be recognizable afterward, right? Please, pretty please?)

And so I too slip into the "fantasy of transcending death." Which is why Shvana's view of melting into the All, like rejoining the Zen river or the Hindu Self, is so challenging. "Becoming part of it" would look a whole lot more attractive if I could be me at the same time. No loss, just gain.

And yet what did Bittner say about that droplet of water at the end of its long fall—"It doesn't lose anything; it gains the river"?

~

SOME PEOPLE, PERHAPS most of us, at some point in our lives feel a moment or an hour of oneness with all that is. For some the experience catches them unawares. An expanding glow overtakes them, ecstatic yet unremarkable as the sun slipping above the horizon at dawn. For others it happens when they are sleep deprived and searching, stretched to the limits with troubles in work or love.

I am of the latter type, more prone to revelations of oneness when I am at my wits' end. Which is why I'm a bit surprised that although plenty of crises have taken place in the past twenty years, I have not had a mystical experience in this sense since I was thirty-four and my marriage was on the rocks.

During that time I was desperate with grief. Life was disintegrating, and I was powerless to put the pieces back together. Having been sick already for a year and a half, I lacked coherence within myself. My insides felt ripped apart. Repressed emotions were surfacing, making me feel crazy and unglued. I was lonely and panic-stricken, sure that a step in any direction would mean worse grief and isolation. Many days I spent weeping.

In this desperate state I consulted a therapist, who suggested some hypnosis. We would visit one of the rooms of the soul, and inside I would find an image of myself, perhaps a younger version, who would shed light on my present crisis.

She put me into a light trance by leading me down stone steps to an underground hallway. (Even as I relaxed I saw the connections with death.) The hallway was carved from stone and lined with doors. When I was ready, I could choose one of the doors, open it, and enter.

I picked a solid wooden door that creaked open on its hinges. Inside was a small circular room with windows stretching from the floor to the tall ceiling. Warm sunlight streamed in through the windows; gauzy white curtains billowed softly in the summer breeze. It was a room meant for sitting on the floor—carpeted, with a low table in the center.

I sat down at the table and waited for something to happen. Nothing did. The room was empty. No images of myself appeared. All was simply quiet, warm, and soothing.

The therapist cast about for what should happen next. "Is anyone there?" she prompted gently.

No, though the room was filled with a sense of presence.

"Perhaps you could ask one presence to become clearer."

Instantly something shifted. I was on my feet, enveloped in a warm, powerful, full-body hug. Opening my eyes in the sunny room, I saw a being of gold-white light, complete with large wings covered in beautiful white feathers. I was being hugged by an angel. It was like being embraced at once by everyone I'd ever loved—parents and brother and friends and lovers and relatives—and all the dear ones yet to come in the future. All that love, all at once. I was held within it, weeping.

At last, from far away, I heard the therapist reminding me gently that I could return to that timeless place whenever I wished, but for now I needed to say good-bye and come back to the present. She led me back up the steps to the light of the everyday, and slowly I opened my eyes.

The world was glowing. Every object shimmered in a dazzling halo of light.

I drove home carefully, afraid the world would fade. It didn't. At home I sat on the back deck, pivoting slowly to look at each familiar object. The towering redwood in the neighbor's backyard—shimmering. The wooden benches on which I was sitting—aglow. Each plant in the garden—shining. The car in the driveway, even the driveway itself—all glowing from within with a light that poured outward. I looked down. My own hands, each finger, each limb was ablaze with brightness.

And the feeling—it was ineffable, though like others before me, I will try to describe it anyway. The glowing world was alive—with laughter and fun, and most of all with love. The brightness itself was loving. Each thing radiated love, shouted its affection. The love dazzled but soothed as well, bathing each part in joyous companionship. All were suffused with cheerful warmth, held within the same radiance. At the center of each, knitting us all together, was love.

I had no sense of being me; I had no sense of not being me. The me-ness simply wasn't important. What mattered, the only thing that mattered, was the bright and joyous love. Wonder was the stuff of the universe; love was each molecule. I felt that I was seeing the world as it truly is.

Over the next few hours the brightness slowly faded. A residue of warmth lingered a while longer.

I wish I could say that experiencing the shining world gave me a lasting peace, an uninterrupted sense of wonder. It didn't work that way. I still had to do the wrenching work of divorce; panic was my companion for months on end. Catching a glimpse of heaven did not prevent me from descending months later into complete despair.

But what it did give me was a picture of the world that felt more real, more true, than the one I held most of the time. I had seen this world with my own eyes, had felt its effects in my body. Twenty years later I can still call up those moments and sense their wonder. The experience became a koan, an alternate view of life that invited me forward, tantalizing me with the possibility that if I untangled the knots of my heart, as the Upanishad says, this dazzling love, revealed once in a moment of grace, might become a stable part of daily life.

I'm still working on the untangling. I'm still terrified of losing my self. It takes practice to recall that this love is most likely the place we all come from and where we all go at the end. Shvana's calm acceptance is still a stretch for me, her picture of death as easy and welcome as fog spreading across the ground still a goal more than a realization.

༅

"ON NO SUBJECT are our ideas more warped and pitiable than on death," wrote a thirty-year-old John Muir. "Instead of the sympathy, the friendly union, of life and death so apparent in nature, we are taught that death is an accident, a deplorable punishment for the oldest sin, the arch-enemy of life." Even after a frightening bout with malaria, he affirmed his faith in a natural world where death and life mingle in "beautiful blendings and communions. . . . Death is stingless indeed, and as beautiful as life."

I doubt that I will ever arrive at Muir's view of a "stingless" death. Every death that I have witnessed hurts. And the closer I am to the one dying, the more it hurts.

Yet I agree with Muir that seeing death as punishment and tragedy goes against a world-affirming ethic. When death is an intruder into an otherwise happy paradise, we are wishing for things to be different than they are. Only when death is a natural part of life are we loving the world as it is.

The view of death as intruder goes along with the idea of a fallen world, a world corrupted by original sin. Which makes it tempting to think that a devaluing of the life-and-death processes of Earth began with Christianity. Surely ideas of resurrection and eternal life, which Christianity taught from the beginning, must have eroded trust in Earth.

But history tells a somewhat different story. The very earliest Christianity modeled its hope in a resurrection, as it had modeled many other beliefs, on Judaism. In the early Roman Empire Jews looked forward to the Messiah being resurrected at the end of time. And along with the resurrected Messiah, God would raise all Jews who had suffered at the hands of Romans or any other conquerors—a vindication for their dying too soon. Early Christians found their Messiah in Jesus, and so they looked forward to the final triumph of Christ and the raising of all faithful Christians. The outline for both was the same: at the end of time God would make things right in the world, giving renewed life to all who had suffered or been killed unjustly because of their faith. For the earliest Church, in other words, as for the Jews of the time, resurrection was a political act, a restoring of social justice. Immortality of the soul had nothing to do with it.

A few centuries later, when ideas of personal immortality had spread throughout the Mediterranean world, Christians were again less different from their neighbors than one might think. Pagan Romans—who worshipped the traditional gods of the city, not those of nature—believed in the two routes after death that Virgil had outlined, one toward "outer darkness" and the other toward a land of heavenly light. The gods of the pagans resided above, in a heaven far removed from Earth. For both pagans and Christians of the late empire, living wisely meant recognizing that the things of this Earth are passing. Honors and riches could not compare to the "true nobility" found in eternity. What mattered most was living an upright life so as to be greeted after death by clean white robes and choruses of praise in the world above. The hero's welcome in heaven looked remarkably the same to both of them.

Life on Earth, in other words, had a penultimate quality. The here and now suffered by comparison with the glories of life after death. Perhaps Romans devalued the physical world because they sensed in ways they could not yet name that Rome's days were numbered. Or perhaps they were merely enacting in their spiritual lives the disregard for nature that had marked the empire since its beginning. Whatever the case, with one eye firmly fixed on heaven, both Christians and pagans pursued business as usual, ignoring the environmental problems of lost forests, poor soil, and bad water that were undermining their way of life. To the very end, Romans would focus on life in the city—if not the earthly one, then the heavenly one awaiting them.

But as the empire was folding, one set of people developed a brand-new appreciation for the natural world. Surprisingly, those people were Christians. Even more surprising, they were ascetics, who sought spiritual peace by subjecting themselves to physical hardship. In the fourth and fifth centuries of the common era, as the urban empire was dying, thousands of these men and women abandoned city life to go live as hermits in the wilderness. They left behind the comforts and frictions of society, gave away all they owned, and retreated to wild and open spaces. Living without possessions, free as animals, they often befriended wild creatures, talking to lions or walking with gazelles.

Some ate only bread and water or deprived themselves of sleep; others punished erotic thoughts with self-inflicted violence. They tried to transcend death by denying the needs of their own bodies. Yet at the same time they often expressed rapturous love for the body of the Earth—an enthusiasm for wilderness that had no precedent in Greek and Roman history.

Basil the Great, a fourth-century bishop in what is now Turkey, built a little retreat hut for himself on a hilltop above a swift-flowing stream and exulted in his special place with words that could rival those of any nature writer today. "Even the Isle of Calypso, which Homer evidently admired as a paragon of loveliness, is nothing in comparison with this," he rhapsodized to a friend. The river below, "for a short space billows along the adjacent rock, and then, plunging over it, rolls into a deep whirlpool, affording a most delightful view to me and to every spectator, and abundantly supplying the needs of the inhabitants, for it nurtures an incredible number of fishes in its eddies." He enjoyed "sweet exhalations from the earth," a "multitude of flowers," "lyric birds," and "all kinds of fruits." The place, he reported happily, helped him find tranquillity.

This is hardly the world-denying attitude we might expect of a monk, especially the bishop-monk who organized all the monasteries in his half of the empire. Yet Basil's enthusiasm was shared by many. Christian hermits were the first in Roman history to love wilderness for itself alone. In doing so they became the unlikely ancestors of nature

writers and preservationists—all those, from Thoreau and Whitman and
Muir to Abbey and Tempest Williams and Butterfly Hill, who leave the
city to seek peace in wild places.

⌐

EIGHT HUNDRED YEARS after Basil a different monk walked the hills and
valleys of central Italy, preaching the benefits of living propertyless and
practicing friendliness to animals. As a young man in his midtwenties,
Francis of Assisi had renounced his well-off family and his own partying
past—the troubadour's life of wine and song—to take up a life of begging and camping out of doors. Where before he was upwardly mobile,
now he gloried in having nothing. Where before he sought nobility, now
he saw in poor people the face of the Christ who had called him away
from riches.

Francis counted animals and trees, birds and rivers among his poor
and humble friends. He was famous already in his lifetime for sometimes speaking directly to them as if they could understand. It is said that
birds listened quietly to his voice and wolves ceased raiding towns at his
request.

Near the end of his short life—he lived to be only forty-five—Francis
composed his famous song of creation, the "Canticle of the Creatures."
Still playing the troubadour, he led his band of wandering brothers in singing to God and their nature friends. They called wind, water, fire, and
earth "Brother" or "Sister," the same way they addressed the humble
members of their own community. But they described each element in
words usually used for noble courtiers. Brother Fire was like the most
admired gentleman: "beautiful and lively and robust and powerful." Sister
Water, like the highborn woman, was "very useful and humble and precious and chaste." To his last breath, Francis would upend conventional
ideas of rich and poor.

Yet even as they sang, Francis felt the song was unfinished. It looked like it would remain that way, for his health was fast deteriorating. He grew ill and weak, and swelling filled his arms and legs. His eyes were already blind.

When one of his oldest friends, a doctor, paid a visit, Francis asked point-blank, "Will I recover?"

The friend hesitated. But Francis, unable to put up with pretense, said, "Tell me the truth."

The doctor replied, "At most, some more days."

At this Francis smiled and opened his arms. "She is welcome then, Sister Death," he said.

He marshaled his last energies and dictated one final verse to his song. "Be praised, my Lord, through our Sister Bodily Death," he sang. "Blessed are those whom she finds doing your will."

Near dusk a few days later Francis lay silent and peaceful. During the hour of his passing, a flock of larks descended to the roof of the house and remained there, wheeling and singing.

ONE MORNING a few months after my visit with Shvana, I received a tearful call from Jackie. Shvana was going into spasms. Lee was out teaching a yoga class and couldn't be reached.

Shvana was dying. The feeling of overwhelm in Jackie's voice told me she knew it too. The vet had said to bring Shvana in, and Jackie was readying her for her final ride.

When they arrived at the clinic Shvana was already unconscious. Within minutes she slipped quietly away without an injection. Jackie stayed with her in a silent room, practicing the letting go that seems to be so much harder for us than for the animals. After Lee received the news he drove across town to join them.

Later I asked Lee if he had any inkling of what was happening. "I still hadn't heard from Jackie," he said, "but when I was driving home after class, this peaceful, calm feeling came over me, like I saw everything very clearly—the grass, the pavement, the trees. Almost like I was melting into everything around me."

10

Family

Teach me, is what you should say, and, I am listening. Approach the world as a child seeing it for the first time. Remember wonder. In a word: humility. Then things come to you as they did not when you thought you knew.

—SHARON BUTALA, *PERFECTION OF THE MORNING*

CHANGE AND LOSS were fast on my heels. In the fall of 2003 Anthony bought my share of the treehouse on the hill, and I made plans to leave the Bay Area.

But I had to do it fast. I couldn't bear to keep walking with Sapphire through the canyon every day and returning home to my treehouse nest unless a new life was calling. But what new life was that? I had everything I'd ever wanted right where I was—friends, community, creative projects. Leaving Oakland made no sense, if you didn't count that little dream of having my own name, alone, on a house title.

I felt like I was flying blind—no voice in my ear for guidance, no synchronicities, no revelations of oneness to provide comfort. I was truly on my own, and completely at loose ends. I felt no leaning toward any particular destination, only that it had to be where I could afford a house, and that meant leaving the Bay Area. Finally I picked Grass Valley in the Sierra foothills, a couple of hours away, and began making weekly trips up to scout properties. At least from there I could visit Bay Area friends.

Except now there were fewer of them. Mimi and her husband, Steve, had just moved to New Mexico; others were heading to Maine. The circle of friends I'd laughed and cried and read poems with for most of a decade was scattering.

I would be moving at the coldest, darkest time of year.

⌐

WHEN THE MOVERS walked in that gray, drizzly morning at the end of December, Sapphire knew what to do. She marched to the front door, her eyes on me. When I let her out, she bounded up the steps to the street and asked to be let into the car. In the backseat she curled up and slept throughout the dark and chilly day, barely lifting her head to give me a long sober look as I packed boxes around her—fragile ceramics, frozen food in the cooler, the clothes I would need the next morning.

A frigid dusk was gathering by the time the moving truck was loaded and we were ready to go. I coaxed Sapphire out for one last squat on the ground—she almost refused—and then we were off.

It was three months almost to the day since I'd decided to leave Oakland.

My stomach was taut, the thin fabric at my center threatening to rip. How could I leave the bay tree–lined woods where Sapphire and I had walked every day, the creek and its annual cleanups, the neighborhood where everyone knew my name? And what did I have to look forward to? Not even a home awaited us; weeks of fruitless searching had led to renting at the last minute a tiny cottage that I'd only peered in at through the windows.

In Sacramento the drizzle turned to steady rain, headlights behind me dazzling in droplet-studded mirrors. On the other side of Sacramento the interstate undulated gently upward, and we turned off toward Grass Valley. Soon I was struggling to follow the unmarked two-lane road. Where was the high beam in my car? In all my years in the city I'd never even turned it on. I groped for a button on the steering column and instantly saw a sky of arcing white. The rain had turned to snow.

By the time we reached Grass Valley, three thousand feet above sea level, the air was icy and an inch of snow covered the road. I wound through town and parked in front of what I was pretty sure was the cottage; the dark and the snow and the fact I'd seen it only once left some question. The key, thank goodness, fit. Inside, I was relieved to find that I liked its clean white walls and gleaming wood floor. Best of all was the gas stove in the living room. I fired it up, and Sapphire instantly plopped down in front of it.

Over the next days, already homesick and lonely, I wondered for the umpteenth time if I'd made a good decision. A friend came up for the weekend to keep me company. "What am I *doing* here?" I whimpered. She reassured me with a warm, cryptic smile: "Well, you're *here*, aren't you?"

One thing I noticed about this transition: it had happened smoothly, with a lot less kicking and screaming on my part than usual. I was determined to trust events as they unfolded. If that took gritting my teeth, so be it; teeth can be useful too. I wasn't exactly enjoying myself, but neither was I sinking into depression or despair. I was bending myself to circumstances not of my choosing, losing a house and a creek and a neighborhood I loved in a city I could no longer afford. Yet much of the time I remained curious about what might happen next. Often I could still find the happiness, the *yes*. It was a first.

The question remained, though: Why was I here?

⌁

JANUARY PASSED IN a fog of work, loneliness, and crisscrossing the county with my realtor in the front seat and Sapphire in the back. The housing market was at its yearly nadir, and the weeks crawled by, excruciating, slow. My tiny cottage was packed to the gills with unopened boxes, and they would have to stay that way until I found us a house to buy.

In town I felt invisible. What was it like to have someone know my name? One night I woke out of a dream in which I could not find a corner to call my own. I was bellowing, "Help! Help! Help!" Sapphire raised her head from my side, looked at me balefully, then slunk down off the bed to spend the rest of the night on the sofa. The next night it was Sapphire who cried out in her sleep, a whining dream-bark.

In early February I put an offer on a decrepit but promising house five miles outside of town. It sat among spreading oaks and a gurgling waterway and lichen-frosted rocks, where I longed to sit and sip tea and watch birds. But I was going to have to get through a shaky escrow first. The home inspector, known for his thoroughness, uncovered a few spots of mold near the washer and advised me to have it checked by a lab. Already weary of problems, I let the mold go.

One quiet evening a few weeks later, after too much television for company, I decided to check up on my friend Tim in Washington, DC. Tim and I had been college friends—well, more than that: two teenagers attracted to each other but too overwhelmed by the connection to explore it. After college we'd married other people and lived our lives on opposite coasts, falling out of touch for more than a decade. A couple years before, about the time Anthony and I were ending, Tim had looked me up in the college alumni directory and sent a letter to catch me up on his life. The day I picked that envelope out of the mailbox, my heart jumped to see his familiar handwriting.

I was surprised—and not surprised—to find parallels with my own life: Divorce. Meditation. Alternative healing. A house in an urban area. Of the group of kids who'd hung out together at that rural Midwestern college a quarter-century earlier, who could have guessed that Tim and I would end up most like each other? I remembered a line from Sweet Honey in the Rock: "Why, your life had plowed the row right next to mine."

Since that letter, Tim and I had exchanged sporadic emails. When, in recent months, he had bought a sporty red car, ditched his glasses through laser surgery, and learned Portuguese to sweet-talk women he was meeting on Brazil's Match.com, I'd teased him unmercifully about his midlife crisis. And when he traveled to Brazil for a month to meet them, I applauded; he was no longer the stoic and stodgy boy I'd known decades earlier.

But now that he was back from his trip, he said, he was even more restless than before. I filled him in on my dismal winter. Finally we got around to inviting each other to our respective homes. I hoped by the end of the month I'd actually have one.

A few days later I headed to a downtown café with my laptop and was sitting outside in a warm splash of early-spring sun when I spotted a familiar face. A familiar face!—one of the first in this town. It was the home inspector, who urged me again to get that mold checked. This time he wrote down the name and number of a lab in Sacramento, and this time I followed through.

The mold report came back toxic, and three days before close of escrow I backed out of the deal. Now Sapphire and I would have to spend even more months in the cramped cottage. Disappointed, I schemed to get out of town. A frequent-flyer ticket—where was the farthest I could go with it? I emailed Tim: "My house deal fell through. I'm bummed. When do the cherry trees bloom in DC?" Less than twenty-four hours later came his reply: "You're invited!"

I flew to DC at the end of March for a long, rainy weekend that turned into a lot more than I'd expected. Tim and I talked for hours, and the delight between us burgeoned into passion. How could I have known that his hands, so eloquent, were the ones I'd been yearning for? His bed the one place I could sleep undisturbed through the sleepless hours before dawn?

By the end of the weekend we knew we would have to rearrange our lives to make room for this new-old love. When we finally visited the cherry trees under a freshly blue sky, the flowers were at their brightest, a fringe of pink lace hemming a soft gray lake. We floated among the trees as drunk as the bees gathering hungrily at their blossoms.

$$\curvearrowleft$$

So I "at last came fully into the ease / and the joy of that place, / all my lost ones returning." Wendell Berry's poem of death and new life had sent me into fits of weeping when I'd found it nearly a decade before. Now I was on the other side of loss, enjoying a reunion that quickened every possible fiber of joy. It was a coming together on a grand scale, of more than just two persons.

Of course, at that level too it was a miracle, for I'd been in love with Tim decades before. From our first semester of college we'd felt the hum of gladness near the other, the feeling of being known and understood. But the teenage Tim wasn't ready for love, at least not with a woman who had the same name as his mother, and he proceeded to push me away then pull me close, over and over again. "I gave you whiplash," Tim said now.

At the time I suffered—lord, I suffered!—but put up with this treatment for three long years before finally giving up. Then forever after I cringed about my twenty-year-old self's inability to get a grip. Tim became the symbol of all my bad choices in love. It didn't help that years after I was no longer thinking about him, he tended to show up in my nighttime dreams.

But now here he was, a peaceful heart who had done plenty of growing-up work, ready to love and be loved. I gulped the ecstatic brew—all the intensity of repressed teenage hormones mixed with the delight of a lively, lifelong connection. I felt an astounding sense of completion, as if all the twists and turns in life, all the dead ends and disappointments, had been for this. "Once or twice in a lifetime," said a line in the book I was editing at the time, "we grasp how all that happened before has led us to this very moment."

A new picture of my life was emerging. The years past were being stitched into an undreamed-of design, and day by day I watched it take on a radiant new shape.

~

IN MY EARLY thirties I made two big decisions, one in work and the other in love. Both choices, in the eyes of many, looked foolish. In work I decided to finish a doctoral program despite knowing that the daily rigors of classroom teaching were too extroverted for me. I simply wanted to research and write, to be part of the conversations made possible by the degree.

In the middle of that doctoral program, sick with a chronic illness, I ended my marriage. To others, the relationship looked solid, but I needed something different—a spark, a livelier intimacy, I didn't know what.

So I said yes to the restless itch and fought for the longings that made no sense. Against all odds I followed what was most unique and creative within, hoping it might find a home in the outer world. But there were no guarantees. Joining inner and outer did help me recover from illness, yet it didn't ensure health or success. It merely threw open the door to them. Outcomes were always uncertain; they depended not just on me but also on the millions of others sharing this life stage. It was never clear if the scenes we were improvising together would ever make sense.

Both decisions in my thirties had required sacrifice—years of loss, a decade of tuition, unstinting work and drudgery and loneliness. They were unsupportable decisions, made on faith—and in what? In some

vision that what is within us is not different from what is outside us—
that nature, both inner and outer, is true and good and worthy of trust.
It was a vision as ephemeral as the faintest star in the night sky: gaze
directly at it and it disappears, but look off to the side and you might
catch a glimpse.

Yet even before I visited Tim, the risks had been paying off. I was using my degree now to teach in the graduate programs of the well-regarded
Prescott College, where I mentored students one-on-one, the learning style
I'd always liked best. Conversations were deep and stimulating, relationships with students rewarding. Colleagues at the school had turned into
friends; every trip to Prescott was like visiting my second home. It was
better than the classroom.

And though I'd had many ups and downs in love, the years of exploring had been rich, sometimes thrilling, and never boring. And now they
had led me to the lover and partner of my dreams. Life seemed to be saying
in every way possible that loss is followed by return, that renewal of life is
the most natural thing in the world.

꙳

MOST HAPPILY-EVER-AFTER STORIES end here: the lovers ecstatic, together at last. We all root for it—the triumph of love against all odds. If
two people want each other, we want them to be together. We want it so
badly that American couples shell out, on average, $24,000 for a wedding
ceremony, hoping it will be the happiest day of their lives.

There's a desperation to the wedding business—as if two people finding each other is a miracle so rare it deserves our most lavish attention. We
seem convinced that love matters little in the scheme of the universe, for
when our most personal desires do get fulfilled, we act as if we have met
up with the extraordinary. As if welcoming love were not the most normal,
animal-like, natural thing to do in the world.

I am all for the joyful surprise—and getting together with Tim was the most joyful and surprising event of my life. We were drunk with amazement and bliss. But I am also aware that part of my joy was the shock of the new thought born from these events: that love in fact *does* get fulfilled. That connection is not only possible but *normal*, built into the nature of things.

I think what people may be searching for in the romantic ending is some hope about nature itself. The ferocity of our commitment to happily ever after might have to do with the even more ferocious tragedy and despair built into our story of nature—into our evolutionary tale of natural selection by competition and into our dog-eat-dog economic life. If nature rewards only scrambling to the top, if at the heart of things the heart doesn't matter and in the world of commerce affection and empathy have no purchase, then love will be segregated to the private sphere, where its most obvious power becomes throwing two people into each other's arms. We celebrate the couple because we are so reluctant to celebrate love elsewhere. Much of the time we can't even see it elsewhere. We've disallowed it from most of our public relationships, including our relationship with the land. "Our lack of intimacy with each other is in direct proportion to our lack of intimacy with the land," writes Terry Tempest Williams. "We have taken our love inside and abandoned the wild."

The death of nature meant, at heart, the death of empathy for other creatures, including human beings. It meant death of the view that connection is stitched into the heart of things, that love is the fabric of the universe. Of course to say this is a statement of faith; of course it cannot be proven empirically. But then, neither can the story we have lived with ever since—of the Earth as dead and connections severed.

When Tim and I came together, I felt reverberations of that alternate story: spirit and matter, love and need, are not separate. I'd gambled long before that they were faces of the same reality, but now I was experiencing it. I was being initiated into ecstasy.

I marveled at the simplicity of the bliss: all those years of searching for purpose, when all along it was simply about connection. Getting relationships right was turning out to be the meaning of my life.

⌐

I FLEW HOME to California, floating, astonished. That spring Sapphire too seemed newly energized. She loved the freshness of the air in the mountains. I often caught her sitting upright in the yard, face uplifted to the tips of the conifers high above and nose bobbing as if she could catch a breeze bouncing from tree to tree. And now that spring was arriving—outdoors as well as in my life—she too was ecstatic. After breakfast each morning she nudged me out the door and waited for me to throw a stick, then grabbed it and played keep-away until I was laughing aloud.

And the last thing every night before going to bed, I lay on the sofa with the phone pressed to my ear, getting to know, again, this person on the other side of the country whom I seemed destined never to forget.

Tim and I talked about our lives during the years we'd had no contact— the relationships, the divorces, the jobs, degrees, spiritual practices. We talked about what each of us wanted in life, and where our lives might flow together. Cautiously I ventured that I didn't think I could ever live east of the Mississippi again and was relieved to hear Tim say that was no problem. Night by night we offered our truths to each other and found them accepted. I was astounded to find that this person knew me—and loved me with no question. Love was simpler than lists of qualities and possible compatibilities, simpler than getting to know another person over months and years (we'd done that decades ago). This was love as a preexisting condition—as simple as opening up to it and enjoying each other.

Sapphire, who'd been trying for years to get this across to me, curled up on the sofa at my feet, her eyelids darting and paws twitching in some faraway dream-chase.

⌒

TWO WEEKS LATER my realtor called: "I'm standing outside a little yellow cottage. I think you better come see this one."

Sapphire and I met her at the gate. In the garden, brick walkways meandered under jasmine-hung arches. Small ponds with rock-lined fountains dotted the yard. The whole was shaded by massive incense cedars, their straight trunks reaching a hundred feet toward the sky. Inside the pine-paneled cottage, there was only one bedroom, but the shower was built for two. It was a honeymoon cottage, and it was ours. Sapphire bounded through the garden, disappearing at the far end to sniff at the fence.

When Tim flew out a few days later for another rapturous weekend, we toured the house together, and when he visited again a month after that, it was to help me move in. We burned dried sage to smudge the cottage room by room and planned where his grandfather's hutch would go. Weeks later, only four months to the day after my trip to see the cherry trees, Tim had packed up his life in DC and arrived in the Sierras to join me.

Later he would say, "I wouldn't have been so ready to relocate if you hadn't just done it yourself." We were meeting on shared ground, brand-new to both of us.

⌒

THE PARTNER OF my dreams arrived with something—or someone— Sapphire didn't expect: an elder cat with soft gray striping and enormous gold-green eyes. Sapphire had her own rules about cats: outside they were fair game for the chase, but indoors she had to leave them alone. It was not something I'd had to teach her. She once weaseled her way past the screen door into the house of a friend with three cats, a friend who was not about to let a dog indoors. Then she lay serenely in the middle of the living room floor, slowly turning her face completely away from each cat who crept close to sniff her.

But she was puzzled by Brio. Some days after he arrived, she started wondering when he would leave. I tried to explain that he and Tim belonged together. *They have a years-long connection, like you and me,* I sent her silently, along with a picture of Tim holding Brio in his heart. Sapphire looked at me blankly: *Why would anyone want to do THAT?*

Since Brio was a visitor, Sapphire didn't have to take him seriously, so she tried to ignore him. This was easy during the day, for Brio was sorting out his cross-country move by sleeping quietly in the dark on chairs pushed under the table, but at night it was a different story. At night he prowled the house and woke with the dawn, ordering the rest of us up with a huge meow. And because he was old and completely deaf, his yowl could wake the dead. Every night it was a shock, first at six in the morning, then a few weeks later at five, then at four.

Something had to be done. Of course Sapphire had been trying to tell Tim this from the start. Although she was eleven now and also nearly deaf, with the first yowl of the night she would spring from her bed on the floor, trot around to Tim's side of the bed, and violently nudge whatever part of him she could reach with her snout: *Go take care of your cat!* It wasn't working. Every night the cat still yowled.

I suspected that Brio just wanted to cuddle in the bedroom with us, though my allergy to cats would not allow so much proximity. But then I heard that aging cats, even the bedroom cuddlers, often yowl in the night. Sometimes they have achy joints (vitamins or supplements may help), or their sleeping area may not be warm enough for older bones.

So when Brio's yowls started at two in the morning, we took action. We created a night sanctuary for him filled with his special things. In the bathroom we placed his water bowl, a soft sleeping pillow like a small throne, a safe heat source, and his litter box. He might have preferred cuddling with us, but having his retreat room did the trick. Before long Brio was sleeping through the night, and so were we.

Sapphire next took issue with Brio over the dinner plates. I'd always put mine on the floor for Sapphire to lick up the juices, and soon Brio decided to join her. As soon as he approached the plate, Sapphire would issue a small, polite growl to let him know he was not welcome. But the deaf Brio paid her no mind.

Sapphire, frustrated by this socially dim creature, began sending him baleful looks throughout the day. About this time, when our friend Jackie asked her about the new members of the family, Sapphire said she thought Tim was okay, but the cat was a mistake.

~

SAPPHIRE WASN'T THE only one having a hard time. Being invaded by Tim, Brio, and their belongings was tipping my scale of change all the way toward unmanageable. It was less than a year since I'd decided to leave Oakland, and in that time I'd moved twice, fallen madly in love, become a new homeowner, and moved into a cottage of less than a thousand square feet with a new partner, enough furniture to fill two home offices, and one dog and one cat.

It was what people call an adjustment period when what they really mean is they wonder how the hell they got themselves into such a mess. Tim and I still enjoyed moments of ecstasy, but they were much rarer, for the stress was activating hormonal changes. I was forty-seven, with hot flashes, irritability, and a sinking libido, just when the love of my life had arrived for good. (Lucky for me, he didn't seem fazed by any of it.) Matches may be made in heaven—ours certainly felt that way—but they have to be worked out on Earth, and the process suddenly looked more complicated than I'd expected.

I missed my old life terribly, and my old friends; the normal process of grieving for what I'd left had been interrupted in the spring with the joy of rediscovering Tim. I felt sorrow also for the years Tim and I hadn't spent together as well as, inexplicably, for the new life we were taking up. About that time a friend said something wise: any new relationship may bring up feelings of grief as we let go of what is possible in order to embrace what is real. It's okay, even normal, in the midst of great happiness to feel great loss.

⌐

LOVE AND LOSS—must they always go together? I was aware that Tim's and my great happiness had been made possible by great losses; each of us throughout our lives had chosen to step away from the familiar and toward the uncertain. The Buddhist teacher Pema Chödrön says, "Only to the extent that we expose ourselves over and over to annihilation can that which is indestructible be found in us." But I must have expected that cycle to be over once I was reunited with the love of my life. Silly me. Things fall apart, says Chödrön, then come together, only to fall apart again.

Welcome each guest at the door, wrote the beloved Persian poet Rumi, even the unpleasant ones like irritability, frustration, grief.

> *Welcome and entertain them all!*
> *Even if they're a crowd of sorrows,*
> *who violently sweep your house*
> *empty of its furniture,*
> *still, treat each guest honorably.*
> *He may be clearing you out*
> *for some new delight.*

I remembered the oldest written tale of a clearing-out, the four-thousand-year-old story of Inanna, Queen of Heaven and Earth. Though Inanna possessed every power in the land of light, she "opened her ear to the great below," the Sumerian text says. She heard a call; she sensed a challenge. She prepared to journey to the land of shadows. By her own strength she would do what no one before her had done: she would return from the Underworld.

So Inanna collected all her divine powers. She donned her royal gown then outlined her eyes in mascara and put on her sexiest breastplate—the one, according to the text, that says, "Come, man, come." Oh, she was hot! She was, after all, Queen of Love and Life and Growing Things. She hung Earth's finest around her neck—precious beads of lapis lazuli—and on her finger slipped her royal gold ring. She picked up her scepter of lapis and her judge's gavel, symbol of her power to decide the fate of others. Death would not dare touch the divine Inanna.

Armored behind her beauty, life, and power, she set out on her journey.

At the gate to the Underworld she rapped loudly. "Let me in! Don't you know who I am?" The gate opened, but then something unexpected happened.

"Give me your scepter," demanded the doorman. "You cannot proceed unless you leave it behind." Inanna protested. "Sorry, Lady," the doorman said. "Those are the rules of the Underworld." Inanna complied. At least she still had the rest of her powers.

She approached the second gate. There the doorman demanded her gavel. This power too she gave up. At the next gates the scene was repeated. Inanna was required to shed her ring, her necklace, her breastplate. Then her robe too was stripped from her.

Finally she arrived at the Underworld—powerless, naked, alone. And there, the text says, she was killed and her corpse was hung on a hook. Not a pretty end for the queen who had everything. And when she died, the Earth also died. Animals lost their fertility, plants could no longer produce seeds and fruit. People lost interest in sex.

Inanna learned a terrible lesson: by her own strength she could not return to life. All her powers crumbled in the face of death. Only when her friends stepped in to help could she leave the Underworld and return to Earth—and then only for six months of the year, and only by sending someone to the Underworld to replace her.

Winter as well as summer would have its season on Earth. New life would be born out of darkness and loss.

~

EVEN AS WE came together as a family, one of us began to leave. Sapphire had cancer, a tumor on her anal gland, a diagnosis we'd received just before leaving Oakland. In the vet's office I wrapped my arms around her neck to steady her while the aspirating needle probed unspeakable places. Vets often asked me to secure her head away from their hands, but with Sapphire it was never necessary; she would have bitten herself before biting any human. On the examining table she usually turned glazed eyes toward a far corner of the ceiling as if dissociating. That day she turned her eyes over her shoulder toward me then yelped and whined.

This was the second cancer in two years. She'd had mast cell tumors removed from her hind legs the year before, and now oncologists found metastasized lumps in her lower abdomen. In three to six months, they said, the anal tumors would prevent her from relieving herself. It was the fall of 2004; Sapphire was eleven years old.

I was heartbroken. Unshed tears clogged my sinuses, yet I couldn't bring myself to let them flow, not while Sapphire was still here, lively as always. Every afternoon she would roust me from the computer with a mighty shove to my forearm with her snout. It was time for our afternoon walk. Soon after I sat back down at my desk to finish up odds and ends, another shove, this time for dinner. Clearly nothing happened on time without her help.

The oncologists were adamant about chemo, but I couldn't see the point. Why prolong the compromised portion of her life? The treatment would not eradicate the cancer, they said, only slow it by twelve or eighteen months. In that time, I knew, Sapphire would have to endure more medical visits, including the exam she had come to hate more than anything: the finger up her butt. *Why do they like to do that?* she complained. I couldn't bear to enter another veterinary office and watch her slink behind chairs to press her rear end tight into a corner of the room, her eyes like those of an abused child.

I decided against the chemo. Instead, we would sweeten our remaining time together—more walks, more treats, more car rides. I began cultivating a moment-to-moment awareness of her, appreciating our connection. I bought her a new dog bed and placed it near my desk. She'd always chosen to lie across the room from me while I worked, but now she settled happily into more closeness. On the sofa she curled up next to me; when I sat at the table she leaned lightly against my leg. I would look up from my work, spot her, and feel content.

I had Chuck do a medicine ceremony for her, and her demeanor suddenly lightened, as if she'd figured out how to step outside the circle of my worry. She resumed playing her favorite game: picking from the garden a rock twice the size of her head, pushing and chasing it, rooting in the dirt, and barking savagely. When Tim rode his new mountain bike through the woods beside our house, Sapphire took off after him, flying headlong down the hill with a burst of her old speed.

～

OVER THE WINTER the four of us settled slowly into our new life. Day-to-day irritations could not erode the feeling of homecoming that I enjoyed sitting next to Tim by day and lying beside him at night. "We've got it bad for each other," I sighed happily to a friend. Tim learned to confine his papers and pens to one small desktop; I learned to avert my eyes from messes anywhere else.

Brio and Sapphire too were coexisting peaceably enough. Brio settled into his chosen spot on the loveseat, looking like the small king he became, and Sapphire resigned herself to living with a cat. But only barely.

As the months passed, Brio came to adore Sapphire. After all, she was a cat's best hope in a dog—a calm canine who gave him a wide berth. He liked her so much, in fact, that he worked hard to win her heart, trying to cuddle up with her if Tim's or my lap was not available. But he would get nowhere unless she was asleep. Then he crept up to her where she sprawled on the sofa and stealthily curled up next to her head. Sapphire would wake to find his small butt close to her nose, and her head would jerk up and seek out my face, her eyes full of reproach: *Why do you allow this?* Quickly she'd jump down and head for a different seat.

But over months Sapphire too softened. When, after dinner, we put a plate down on the floor for her to lick, she no longer growled as Brio approached. Tim and I laughed to see them side by side concentrating, one bristly cat tongue noisily scraping, one long dog tongue smoothly lapping. Eventually she even tolerated Brio's presence on the sofa. We would come home to find them curled up inches away from each other, sometimes even bumped up together, snoozing.

One night after Tim returned from choral practice and I from ceramics class, we mused over chardonnay and snacks about our younger selves twenty-five years before.

"It would have been nice to know then what I know now," I said. To Tim's questioning look I added, "I would have changed my life sooner."

"What do you know now?" he prompted. He'd always asked the right questions.

I thought a moment, the wine warming my belly. "Oh, that things always work out, one way or another. That you don't have to stay in an unhappy situation—that happiness is possible.

"And most of all—I don't know if this works for everyone, but it's sure been true for me—that we often get a second chance."

⌒

BY SUMMER SAPPHIRE was clearly slowing down. We were entering that room of impotence, watching the loved one worsen, then rally, then worsen again. One night when Sapphire was on the downslope I had a dream of traveling around the country, crisscrossing the landscape. I saw grasses and rolling hills and houses on steep, sunny mountainsides. From time to time I met up with Sapphire, who also was traveling around the country, riding on a tiny log platform that whizzed along just above the ground. She was unbelievably happy to be covering the ground at breakneck speed.

I woke up with a start and leaned over the bed. Bending low, I could hear: she was still breathing. But in the days after, she began to seem dreamy, not quite present, as if some part of her was already traveling.

⌒

WHEN I WAS seven our dog died, overdosed with anesthetic while being spayed. Mitzy was not an indoor dog—Mom would never have allowed it—but I felt close to her anyway. I was inconsolable. Mitzy was gone, and there was nothing I could do about it. She was dead. Not coming back. Ever.

I wanted to scream, hit people, take revenge—anything not to feel so powerless. But I was too old for tantrums and too young for revenge, so I turned the grief inward, wearing sadness like a cloak of midnight. These days we recognize depression in children, but that was the midsixties and I was mostly on my own. My mother tried to console me, but I heard her unspoken words: this grief was way too big. Mitzy was only a dog, after all, and we could always get another.

It was not my first experience with death. When I was three my grandfather died after lying sick in his bedroom for days. Even at three I understood that I would never see him again, but the thought was not painful. He was very old and not a big part of my life.

Mitzy was different. Mitzy and I shared life. When she was gone, I knew she could not be replaced. Time did soothe the pain, and I did enjoy our next dog, though I did not bond with any of our new dogs in quite the same way.

Now, as I watched Sapphire decline, I remembered that feeling of hitting the wall, at seven and again at thirty-five. Death couldn't be reasoned with. It was rigid, authoritarian, like my parents had been. Wielding its power like a tyrant, taking things away with no explanation. Absolute. Harsh.

Would all I had learned about trusting life help me now trust death?

One morning at the end of my yoga practice I rested in shavasana on the sofa. Turning my attention inward, I saw myself approaching the wall. This time, was there a way to be present, to meet it without fighting so hard, without crashing full tilt?

An answer came, slipping in around the edges: *Be grateful for what is. All of it, in this moment.* I tried it out, tasted it bit by bit. I felt my gratitude reach around our home—around Tim, around the cat, around Sapphire, the beautiful garden, the lives we were building in this town. Yes, all of that I could feel grateful for.

My attention stretched to Sapphire's coming end. Could I be grateful for that too?

No way. Absolutely not.

But what if I could? What if it were possible to feel grateful, not just for her life, but also for her death?

I felt my resistance build. For once I chose to just keep breathing, watching.

Suddenly, there was a pinpoint, an entrance. For a split second I could hold even her death in arms of gratitude. Or imagine holding it.

But it was only an instant, and it passed. Then I snapped back to feeling resistant.

But that instant—so peaceful!—not having to fight what is, even when "what is" is death. In that moment, so much happiness available!

I heard Sapphire get up from her dog bed across the room, felt her crawl stiffly up on the sofa and settle against my legs, her head across my ankles. Her flesh was warm, her fur soft against my skin.

We breathed together, our eyes closed.

⁓

IN MID-SEPTEMBER SAPPHIRE began to have trouble relieving herself. It was nearly a year since the oncologist had said this would happen in only a few months' time. Still, she trotted around the block on our afternoon walks looking happy.

Every week I took her to the holistic vet—acupuncture for pain and discomfort, Chinese herbs for energy. Dr. Peggy marveled at Sapphire's composure. During each treatment Sapphire stood quietly, watching. On one of our last appointments, after all the acupuncture needles were in, Peggy said, tears brimming, "She's such an amazing dog." I was glad she could see it too.

"It's her communication," Peggy added. "I find it so easy to treat her. She's so clear about what she wants and doesn't want. That direct look—yes, that spot is fine today. No, that one isn't. I can work *with* her. I've learned a lot from her."

"Like—?"

"To listen to the animals. To take my lead from them. Not to impose my will on them. That's what my teacher always used to tell me. 'Just listen. They'll let you know what they need.'

"And she's kind," Peggy added. "This thought keeps occurring, that she just wanted to make sure you and Tim are going to be okay together. Most dogs wouldn't have wanted to share you. But she wanted you to be happy. It's very unselfish of her, and kind."

Peggy smiled, beginning to remove the needles. "She's a wise being."

༄

ONE MORNING AT the end of September Dr. Peggy came to the house, injections in her bag. I was fairly clear, free of tears, as I had intended to be, ready to be present to Sapphire.

I had explained to Sapphire what would happen. The needle would go in, and then—what? I had no idea.

The needle went in and stung her. She yelped and tried to rise and flee. I stayed with her, holding, comforting.

Then one last, surprised glance over her shoulder at me—what did that glance mean?—and she sank quietly to the floor. Her breathing faded. I waited, holding back the tears. How would I know when it was over?

Suddenly, for an instant, she appeared in my mind's eye. She was jumping and slinging herself in midair, ecstatic, free.

Then she was gone.

༄

WE LAID HER body out on a mattress on the floor. It took most of the day to cool. Brio approached her, sniffed gently, then curled up inches away and slept.

Grief overwhelmed me. Every few hours I found myself on hands and knees on the floor, sobbing. Tim stayed near me, holding me when it helped, sitting nearby when it didn't.

Where had she gone?

It's closer than you think. This sentence kept going through my mind, sliding into awareness even past the raw pain. *It's closer than you think.*

~

OVER THE NEXT days we found a new rhythm. I often woke in the early morning, grief-panic propelling me out of bed. Instantly Tim would wake and rise to be with me. (This was a miracle of love; he is not in any way a morning person.) We would slip into our shoes and head to the woods for a rapid hike to work off the panic.

One morning before dawn I had a dream. I was riding on a passenger train, watching up ahead while the engine in slow motion slid off the tracks. I was pissed off; this was the same place it had derailed before. I slid out of bed, and then Tim was up too. While we walked in the woods I told him my dream. "I've been here before," I said. "Feeling a lot of grief, more than my partner could handle. I wonder if it's going to happen again. I'm afraid."

We walked in silence for a moment, then he said quietly, "I'm not going anywhere."

I didn't know this was what I needed to hear. "You're not?"

"I'm not going anywhere," he repeated. I burst into tears as he turned toward me and gathered me into his arms.

11

Brio

Find god in rhododendrons and rocks,
passers-by, your cat.
Pare your beliefs, your Absolutes.
Make it simple; make it clean.

—SHERI HOSTETLER, FROM "INSTRUCTIONS"

B RIO STEPPED DETERMINEDLY onto my lap and sank into place. He was doing that a lot now that Sapphire was gone, and I welcomed it. In those first days after her death, when my grief burned white-hot, I was surprised to find that Brio provided a corner of cool shade, a tiny respite. Someone with fur and four paws lived in our house, and he helped, if only a little.

Over the winter I healed little by little from Sapphire's loss. But though I was laughing again, something was not quite right. I was staring down fifty, and I was dragging: How could I again find the *yes*? (And why is it so hard to trust the seasons of closing as much as those of opening?)

Winterhawk the animal communicator paid us a visit and instantly fell to adoring Brio. He grew softer in her presence, more peaceful and content.

273

Before she had been in the house ten minutes she turned to me and said, "Brio wonders when you're going to write *his* chapter."

It took me a second to comprehend; I hadn't even told her yet that I was writing a book. Write about Brio? He was Tim's cat. Tim monitored his moods, prepared his food, adjusted his supplements, made sure he was warm or cool enough at night, brushed him every evening, cleaned up his hairballs, mopped up his diarrhea, adjusted his supplements again, and brushed some more. Though I often helped out, especially with brushing, Brio seemed opaque to me. By now, more than a year into our relationship, he was no longer the blank wall he'd been when he first arrived, but I certainly didn't know him well enough to write about him.

A chapter for Brio? It was a new thought.

In May 2006 I was called to Boulder, Colorado, on business, a follow-up trip to my first visit two years earlier, when I'd fallen in love with the town's sunshine and mountains, its university campus and library. But during that earlier visit Tim and I had been newly nestled into our cottage, working hard to adjust to each other and our new home, so instead of setting out for the wide skies of Colorado, we'd redoubled our efforts to settle in under the cedars and pines of Grass Valley. And we'd succeeded. We now had dear friends in town, satisfying lives, and lots of places to hike, including spectacular vistas in the High Sierra, only an hour's drive away from home.

Yet in the week leading up to this second Boulder trip, I found myself getting nervous. What about? It couldn't be the business content of the trip; I had met these colleagues before and knew what my job entailed. Then what was giving me the jitters?

Not until after dinner on the night before I was to leave did the renegade thought occur: *What if I still love Boulder?* The question was shocking. After all that both Tim and I had sacrificed to get to Grass Valley, how could I possibly be thinking about Boulder? But the thought would not go away.

Finally, full of trepidation, I voiced it. "Well, let's think about it," Tim responded slowly. He was less surprised than I'd expected him to be. We sat the whole evening considering this new question. By midnight, we had made a decision. If I still loved Boulder, we would be open to relocating.

The next day Boulder met me with summertime heat and leafy green trees. My hours-long meeting was held on the patio of a house looking out toward the Flatirons, those sandstone megaliths defining the edge of town. Birds twittered in the trees overhead; a deer came to rest in the shade beside the patio.

Yes, I still loved Boulder.

When I returned home Tim and I delved into the question of moving. We talked of it over dinner; we mused about it while petting Brio; we discussed it on wildflower-studded walks in the foothills. At the end of June we made an exploratory trip to Boulder to see if Tim liked it too.

Near the end of our Boulder visit, still unsure if we belonged in this town, we decided to consult a larger wisdom. We would seek out the voice of the stream slicing its way through the center of town. What did Boulder Creek have to say about our move? Parking next to the creek, we turned our separate ways along the creekside path, each of us finding a spot to sit quietly for a while with our own thoughts. The creek in June was deafening, a rushing tide of snow-fed waters that roared its way downhill over rocks. The air next to the creek was fresh and bracing, the sound of the waters invigorating. When we later compared notes, we had come to the same conclusion: the path toward Boulder appeared open.

Returning home, we began making plans to move across the country.

⌒

EVEN BEFORE OUR Grass Valley cottage was listed for sale, our realtor called to say someone would come by to see it early the next morning. We slaved for a twelve-hour day of cleaning so we could head to the high country the next day for a long-overdue hike. Brio could stay at home and sleep as usual while the strangers let themselves in and toured the house. He might not even notice they were there.

Why, after years of living with animals, did I still think like that?

The next morning dawned sunny and dry, promising another July day of record-setting heat. The house looked spotless, but I primped the living room anyway, moving the fan to a less prominent window. This was a charming knotty-pine cottage, after all, and a metal box fan should not be the first thing a buyer saw when stepping inside the front door.

But the window where I placed the fan happened to be the one where Brio liked to sit and gaze at the garden, a north-facing window with a cool and shaded sill. I thought I'd left Brio plenty of room to sit in the window *beside* the fan. Apparently he didn't agree because as I was doing my morning yoga he stalked to the window, paused to look out through the space that was left, then jumped down from the window and stomped past me, his tail rigid.

Foolishly, I ignored him.

Tim and I ate breakfast, packed our bags with sunscreen and water bottles, said good-bye to Brio, and headed to our favorite trail at Grouse Point, an hour up the mountain. At the trailhead the air was clear, the placid waters of Carr Lake jewel-blue. The last time we were here, on Memorial Day, the lake had been sheeted in ice. But now Sierra wildflowers coated the hillsides. We felt jubilant. We might get a buyer before the day was out, and then we could be on to the next adventure. A realtor in Colorado awaited us.

We hiked past Carr and around the bend to sparkling Feeley Lake, then up the hill past granite boulders and pine trees, panting. Halfway up the hill we stopped to catch our breath and glimpsed a streak of brightness darting through pine branches. A male western tanager with his radiant orange-red head and golden body made his leisurely way around the circle of trees once, and then again, calling and twittering, the only sound in an otherwise still forest.

Finally up at Island Lake, we followed the trail to our favorite granite outcropping and sank down on rocks already warm at midmorning. The Sierra lake was deep azure, almost cobalt, ringed by granite hills blazing white in the sun. The lake's tiny island, no more than a few rocks and scraggly pines, gleamed at its center. We sat for only a short while because we needed to get home in time for lunch. Besides, we didn't want Brio to be left alone too long on a day when something unusual had taken place in his house. How would he react?

When we had taken in as much of the brightness as our eyes would allow, we made our way back down the trail. The high country had worked its magic; I felt lighter inside, more ready to laugh again. Tim's face too was relaxed and smiling. We bumped out the dirt road and drove the curving highway to home.

The house was quiet, spotless. Or so it looked. The realtor's card lay on the table.

Tim saw it first: cat shit by the front door, on the heirloom rug woven from his grandmother's sewing scraps. Brio was in his usual spot on the love seat, but he wasn't sleeping. He wasn't even lying down. He was sitting up, glowering. Just behind him, on the low bookshelf that served as his pathway to the favorite window, more cat shit—which wouldn't have been a big deal except it lay on a green silk runner embroidered in gold. Brio was clearly upset. He had sent the message that angry cats often send.

Tim and I sat down with Brio to calm him. What exactly had been so unsettling? Did he hate the people who toured the house? Was it the fan in his favorite window? Yes, of course, the fan was a problem. I moved it back to the original spot, freeing Brio's window. But was there more?

I quieted my mind, making myself available to his world, ready to catch sight of a perspective not quite my own. Finding a clear mind should have been easy today after spending several hours among trees and lakes, but no luck. After only a few seconds I was lost in a memory of Sapphire walking through the house, her one blue eye shining lightly as she sought me out with her gaze. Suddenly I caught myself: *I'm not supposed to be thinking about Sapphire, I'm supposed to be listening to Brio!* Ever the agitated monkey, my mind was always jumping to another tree.

But before I could chide myself further, a different thought stopped me: *What if Sapphire* is *the point?* I stopped the mental stream to consider it. What did Sapphire have to do with Brio's problem? Then, in a flash, another thought: *Sapphire was a dog, a watchdog. A cat doesn't want to do a dog's job!*

Now that Sapphire was gone, Brio was afraid that he would be expected to guard the house as she had done. That morning strangers had entered unannounced, and he'd been helpless to do anything about it. He'd failed at the watchdog job, he'd failed us, and he'd been frightened to boot.

We apologized to Brio. We petted him, telling him we understood. We assured him that we didn't need him to be a watchcat and promised not to put him in that frightening position again.

And we kept our word—by letting Brio know ahead of time what was going to happen. Many other realtors and clients passed through the house that fall. Each time, we left for an hour or three, and Brio stayed home alone. But before we walked out the door, one of us sat down beside him on the love seat and sent him pictures: we were leaving now, some strangers would enter and walk through the house, they had our permission to be there, and we did not expect him to guard anything. "In fact," we told him, "your only job when they arrive is to look beautiful," a job that he of course excelled at.

No more upset-kitty messages awaited us. Every time we stepped in the door after realtors and clients had left, Brio was curled up on the love seat, sleeping peacefully.

~

My FINGERS PAUSE on the keyboard after writing Brio's story, and in that moment Brio slips around the corner of the table, stopping beside my chair and gazing up with his wide golden eyes. I scoot back, and he springs to my lap like water flowing uphill. He sits upright for a moment, his purr a growl that can be heard at the other end of the house, then tilts sideways to fall into my chest, onto my heart. I feel his warmth and softness, then something extra. *I love you.* He is grateful I am finally writing his chapter.

I love you too, Brio, I assure him, stroking his silky gray fur.

~

BRIO'S STRONG WILL is familiar to anyone who lives with a cat. It is unremarkable, expected even. Cat people love to trade stories, and most of their stories showcase the fiercely strong feline will.

But cats are not the only creatures with determination. European-derived societies share the bedrock belief that nature itself is willful, which is why they have spent so much effort trying to control it. Willful nature exists at the edges of awareness, in the negative space between skyscrapers and smartphones and sumptuous fruit displays in markets. All the beating back of forests and deserts, all the pesticides dumped on crops and lawns are a testament to the belief that nature intends to frustrate human efforts to sustain ourselves, that the life we want cannot be had without fighting off that original foe.

The English word *wilderness* comes from the Teutonic and Norse root *will*, meaning self-willed or uncontrollable. As environmental historian Roderick Nash traces it, the older word *willed* merged into *wild*, meaning unruly, like water rolling in a mad boil. A wild person was out of control, ungovernable. Applied to animals (the Old English word *dēor*), a *wildēor* was an animal not under human management. The word *wildēor* shows up in the eighth-century tale of Beowulf to describe ferocious animals living in miserable forests and on unscalable cliffs—out-of-control beasts in a hostile environment. Unmanaged nature is disorderly, contentious.

꒕

WHILE I AM writing Brio's chapter I pick up, for company, a book I've long planned to read. The themes in *Epitaph for a Peach* converge with the themes that are filling my thoughts.

"The moment I step off my farm," writes the California fruit grower David Mas Masumoto, "I enter a world where it seems that everything, life and nature, is regulated and managed." Masumoto has decided to forgo applying pesticides to his peach trees and is struggling with the out-of-control feeling that follows upon giving up chemical warfare.

Yet just beyond the borders of his farm the machinery of control lumbers on—heating and air-conditioning to prevent people from being in contact with the weather, insurance and bureaucracies to protect the public from risk. "In America," he says, "a lack of control implies failure."

I take a half hour from writing to go dig dandelions out of the tiny plot of green in front of my house. While I am on my hands and knees a delivery woman brings a package up the walk. She takes one look at the yard and grins. "I think you're losing," she says.

"Don't we always?" I grin back, aware that I don't stand a chance against the gritty, promiscuous life in the yard—if I make it about winning and losing.

꒕

SEVEN OR EIGHT thousand years ago, when farmers first migrated down from the mountains to the lowlands of Mesopotamia (now Iraq), they found flatlands nestled between two great rivers. There was plenty of water, but for farmers it flowed in all the wrong places—too much here, not enough there. Lands that weren't marshy were arid and dry. Hunter-gatherers lived happily between the rivers, finding plenty of fish, game, and edible plants, but farmers simply had to control the water.

They began modestly enough, cutting short canals through the dikes that the ever-flooding Euphrates had built up along its banks, sending the water flowing into nearby fields. Century by century their efforts became more elaborate, until systems of canals and weirs and levees and dikes crisscrossed the land. The first cities six thousand years ago, then the first empires two thousand years after that, were made possible by controlling water.

Setting the boundaries of water was so crucial that canal maintenance came to be the judge of government. A good ruler was one who cleaned and dredged the canals, a bad ruler one who allowed swamps and deserts to creep again across the fields. Calendars were dated by the digging of new canals, and rulers attached their names to the canals "they" dug. By the time Cicero in the first century BCE was boasting about the dams and dikes built by Romans, the Near Eastern habit of assessing a civilization by how well it controlled water was already thousands of years old.

And of course the act of controlling water played a central role in myth. The creation story of ancient Mesopotamia, called the *Enuma elish*, its written text discovered in the vast libraries of the Assyrian emperor Assurbanipal (600s BCE) but probably recited for centuries before that, tells of the male hero-god Marduk leading a cosmic battle against female Tiamat, the great sea monster of chaos. Marduk prepares his weapons, the evil winds of cyclone and hurricane, and lets them loose in her face. Startled, she opens her mouth and swallows them, and they distend her belly. Marduk shoots an arrow, piercing and killing her. He stands upon her bloated body and splits her open. From one half of her Marduk creates the heavens and from the other half the earth.

The primal chaos of water must be subdued, said the ancient Mesopotamians, before human life can begin. Every New Year they

celebrated the slaughter of Tiamat and the creation of the world by recit-
ing the *Enuma elish* with great solemnity.

⌒

WHILE READING ABOUT ancient Mesopotamia, I learn that the Hebrew
Scriptures (the Christian Old Testament) owe their present form to the
decades-long exile the Jews suffered in Babylon (the next empire after
Assyria) in the 500s BCE. Catastrophe befell the kingdom of Judea when
it refused to pay tribute to the far-off emperor in Mesopotamia. The land
was invaded and the ruling class deported to Babylon.

Exile meant the death of Jewish social institutions, the end of the world
as they knew it. No longer able to trust the stories of the past, the Jews cast
about for something to hold on to. What, in this world of calamities, could
be trusted?

Eventually they found new answers, began to tell new stories, distill-
ing a fresh faith from the troubled brew of exile. Their scriptures are an
anomaly among ancient Near Eastern texts, written not by the ruling class
in Mesopotamia but by the conquered people.

One thing the Jews made clear in the opening chapters of Genesis: creation itself is good. Nature is trustworthy, though it also tends to be wily;
like the serpent, it presents tricky choices. But unlike in the *Enuma elish*,
nature is not raw chaos to be subdued. Every part of nature, the Jews said
emphatically, is good.

On a second point they were just as clear: the God who formed this
good Earth is benevolent. Unlike Marduk, their God did not create through
slaughter. Neither was he to blame for their present predicament. Though
events in the Jews' lifetimes had taken a disastrous turn, it was not God's
fault. About this they were adamant: reality itself can be trusted.

What, then, explained the current tragedy? The fault, said the Jews,
lies in humans, especially in their overreaching. The Jews knew firsthand

what overreaching looked like: faraway kings who demanded tribute without reciprocity; vicious armies who swept in and bludgeoned or starved people into submission then forced survivors to trudge hundreds of miles to become slaves in a foreign land.

Humans tend to make bad choices, said the Jews in exile, in the creation story that became Genesis 3. People often choose arrogance and might, defying the Creator by usurping powers not meant for humans to wield, especially not against one another. And once they eat of this fruit, miserable conditions follow, for such choices drape the land itself in tragedy. Food becomes hard to grow—think of the enslaved field workers in Babylon whom the captive Israelites were forced to join; relations between people sour, with men lording it over women (were they thinking of Marduk and Tiamat when they wrote this line?); and life itself is cursed.

Read in light of a people in exile, the creation stories in Genesis are concerned with cultural trauma, their gaze turned steadily on injustice. The exile in Babylon became the touchstone for a faith that forever after centered on the theme of social justice. Jews of later eras, such as Jesus in the Roman Empire and Marx during the Industrial Revolution, would draw from this foundation to criticize the inequities of their times.

⌁

LIVING WITH BRIO was an exercise in lack of control. I suspect he felt the same way about living with us. We despaired about his iffy digestion, his cranky moods. He despaired whenever we ventured into the garden without him, bellowing his misery in wails that could be heard a block away. In his younger years, when he lived beside busy urban streets, Brio had been an indoor cat, but in Grass Valley I gave him supervised walks in the garden. Every being needs to feel the soil, to touch the grass, to absorb an unfiltered sun. Perhaps he was cranky *because* he'd been an indoor cat.

He lived for our walks. In the garden he stalked silently beside shrub-bery, pranced proudly down brick walkways, sat amazed before birds in grass. After nightfall he streamed quietly through the darkness, trying to elude me; I practiced sensing his motion when I couldn't see or hear him. Was it only my imagination, or was he feeling happier in the rest of his life, less fretful? Indoors he continued walking determinedly onto our laps, set-tling down in satisfied contentment.

Then, just as we were preparing to move to Colorado, Brio began sneezing blood. Abscessed teeth were removed. The sneezing continued. We would walk in the house to a scene of murder, drops of blood spatter-ing the floor.

Tests were inconclusive. Still we kept searching for the cause. If only we could find out what was wrong, surely we could fix it.

AMERICANS ARE LOYAL to the ideal of individual liberty; we resist any limits on our free will. Yet there is a way in which we are, after all, servants of control. We need the world to be rational, not fickle; we swear by cause and effect. Physics professor Victor Mansfield even called it an "idolatry toward determinism and causality." The scientific revolution was built on the notion of an orderly world where the past logically produces the pres-ent, and for more than three centuries a strict determinism ruled Western science. In 1812 the philosopher Laplace wrote that if there were an intel-lect big enough to calculate all the movements of huge planets and tiny at-oms at the same instant, that intellect could know every motion in the past as well as the future. "For such an intelligence nothing would be uncertain, and the future, like the past, would be open to its eyes." His words cut to the heart of the issue: the out-of-control feeling that arises in a precarious world. Scientific laws of cause and effect were aimed toward the exact same goal as the prescientific thinking of previous millennia. Like the spells, rituals, and abstract philosophies before them, laws of causality were intended to banish uncertainty from human life.

Though twentieth-century quantum physics exploded confidence in causality by reasserting the role of chance, most of us continue to resist the idea that events may not be determined by rational, knowable laws—that the universe might act irrationally or, to put it a better way, that the universe itself might have free will. To the end of his life Einstein defied the quantum physics revolution, saying, "I am convinced that God does not throw dice." (To which his friend and colleague, the quantum physicist Niels Bohr, retorted jokingly, "Einstein, stop telling God what to do.") Victor Mansfield adds, "Rationality recoils at the idea that *in principle* a causal explanation cannot be found. This violates the Western devotion to rationalism, the foundation for science and analytical thought in general."

It is a commitment to causality that lies deep in the religious foundations of Western thought.

↝

JOHN CALVIN, BORN just a few years before Martin Luther's Ninety-Five Theses launched the Protestant Reformation in 1517, studied law as a teenager at the best French universities, sitting at the feet of the brightest Renaissance thinkers of his time. True to his lawyerly education, he took it as his life's mission to spell out the legal implications of Christian thought. The churches descended from him—as varied as Presbyterian, Congregational (later United Church of Christ), Baptist, and all stripes of American fundamentalism—share an emphasis on understanding scripture texts correctly and even, as among fundamentalists, on taking scripture as a legal document, a written rule.

Calvin wanted to understand how the new Protestant idea of spiritual autonomy—every church member free to approach God without priests as

go-betweens—jibed with the idea of an all-powerful God. If God is omnipotent, do human beings have any real free will? In this question Calvin
was a true modern, anticipating by nearly two hundred years the question
that became urgent after Isaac Newton: In a strictly deterministic world,
does choice play any true role? The answer is clearly no, if you start with
either an omnipotent God or his proxy, a clockwork universe. But Calvin
wanted to find a different answer. He pressed on, undeterred.

He claimed to get around the problem with the idea of predestination,
which he borrowed from Augustine. God chooses whom to save, Calvin
said, and nothing humans may or may not do can change that choice. God,
in other words, has complete free will, and God's decisions are certainly
just (read: rational) even if they are beyond human understanding. "God's
will is so much the highest rule of righteousness," he wrote, "that whatever
he wills, by the very fact that he wills it, must be considered just." Humans
have free will too, though we use it, he said, as Augustine before him had
said, to inevitably choose sin.

Calvin was a theological prodigy: all this he penned in his magnum
opus at the ripe old age of twenty-six.

Predestination led in turn to the idea that all events are foreordained,
or known and planned by God ahead of time. Which led in turn to the old
seminary joke: What did John Calvin say when he slipped and fell on the
ice? "I'm glad that one's over with!"

Calvin's commitment to a rational (if unfathomable) God lives on in
part in the idea that all events of nature have an explanation. The more
judgmental among Calvin's heirs use natural disasters to point fingers at the
scapegoats du jour—women, feminists, gays and lesbians, Muslims, anyone they consider immoral—a tactic that would have horrified Calvin since
he regarded it as futile, even blasphemous, to inquire after God's ways.
The scapegoaters inherit Calvin's idea that events in nature are divinely
ordained, though it boggles the mind to think that their position evolved

out of the most rational, Renaissance-oriented wing of the Reformation. (They had to borrow as well from the cast-in-stone quality of Newton's laws, among other things, to arrive at such rigidity.) The shadow side of strict causality now becomes clear: if all is knowable, then humans can take it upon themselves to pass moral judgments on others.

I have always loved, for contrast, poet Mary Oliver's take on natural disasters. She writes with calm, open-eyed assurance in her poem "Shadows" that flood and whirlwind are not choosy. No one is out to get us; no cyclone is a divinely guided Uzi sent to blow evildoers away. Nature is radically democratic—an idea preached long ago by Jesus, who said in the Sermon on the Mount that God makes the sun to rise on both evil and good and sends rain on both the just and the unjust (Matthew 5:45). Simply put, disasters hold no hidden moral message. They can't be used to point fingers at others.

~

THE MURDEROUS INTENT of cyclone or flood is precisely what Europeans rejected at the dawn of the modern era in their rush toward rationality. To arrive at a safe and predictable but soulless cosmos, they had to sacrifice much—for starters, the feeling of being surrounded and supported by other beings with whom they might converse or consult, not to mention the sense of meaning that resides in such relationships. If the cosmos was not murderous, it could not be benevolent either, so they jettisoned also the feeling of well-being that comes from living on friendly terms with thunder and lightning, with stately guardian-trees, with life-bestowing springs.

To us in hindsight, their intellectual choices look political more than purely rational; they were struggling to break the power of the medieval Church and install in its stead a new, scientific, elite. And they were trying as well to stamp out livelier, animistic views of nature in their own towns and villages and among the people on their own estates as well as among the foreign peoples they were conquering.

Perhaps this is one reason the Enlightenment's mechanistic view of nature lodged itself so firmly in the Western mind-set—not only because it allowed people to understand, predict, and eventually manipulate their environment; not only because it gave people the comforting sense that all was ordered and tidy; but because it dispelled old specters of superstition and magic that Europeans wanted to think they were leaving behind. Leaders of the Age of Reason in the seventeenth and eighteenth centuries were determined to drive out all remnants of that earlier irrationality— leftover medieval rituals and spells, all magical practices that could not be explained by the new science. The English philosopher Francis Bacon wrote in a book published in 1627 about the awful superstitions and "monstrous imagination" linked to the ancient idea that the world is alive. Those archaic misguided folk believed that "if the world were a living creature, it had a soul and spirit." One can just about see his hands raised in horror. Such thinking opened the door to all sorts of irrational and unrestrained magic. "With these vast and bottomless follies," he clucked, "men have been in part entertained."

The living universe was also the worldview of people who were being conquered and enslaved during Bacon's time. For it must never be forgotten that the scientific revolution took place alongside the voyages of conquistadors and colonists and slave traders and was sometimes directly financed by them. Mastering nature went hand in hand with mastering other people. The two forms of domination were linked through an ideology of a mechanical, inert Earth that, on the one hand, made possible a greater control of nature at home and, on the other, appeared to justify the superiority of English or European ways over those of Irish, African, or American natives who revered the living Earth.

⟳

TODAY A RESIDUAL fear of superstition remains, and it rears its head whenever there is talk of a personal view of the universe, whenever the possibility of an animate world enters the conversation. The question always arises: If the forces of nature are alive, having their own free will, don't our lives inevitably turn to pacifying them? If nature is personal, aren't humans reduced to groveling before other beings, begging them for their good favor? Doesn't faith devolve into superstition?

It's worth investigating for a moment what superstition actually is. The usual definition is "irrational belief about causation," such as a black cat crossing the path and bringing bad luck. Omens like these were emblematic of the culture that early modern Europeans worked hard to eradicate. Would the early moderns have paused for even a moment if they'd known just how far back in Western history such beliefs go?

⟳

IN THE GREAT Assyrian empire of the 600s BCE, centered between the Tigris and Euphrates, reading was godlike. Not limited to deciphering marks on clay tablets, reading extended to following the motions of bodies in the heavens and on Earth. To read was to understand the movements of moon and stars as well as the mirrors of the heavens that one could find by sacrificing an animal such as a sheep and cutting it open to study the positioning of its liver and entrails. And of course reading included all the cuneiform imprints people made in clay to record those happenings. Reading meant understanding signs of every kind; it encompassed what we think of today as astronomy, medicine, history, literature, religious studies, and divination.

The skills of reading made up the schooling of royalty. King Assurbanipal, the fantastically powerful—and brutal—Assyrian emperor in whose library the *Enuma elish* was found, in the 600s BCE described his early education:

> *I learnt the craft of Adapa the sage, the hidden mystery of the scribal art. I used to watch the signs of heaven and earth and to study them in the assembly of the scholars. Together with the able experts of oil-divination, I deliberated upon [the text] "If the liver is a mirror of heaven." . . . I looked at cuneiform signs on stones from before the flood.*

To read was to divine, to commune with the gods.

Talking to gods played such a central role in Assyrian life that a literature of omens took shape over generations and became a fixed part of their canon. There was, in other words, an "omen science." Each succeeding generation of rulers copied lists of omens onto clay tablets, all the young sons learning their "letters" by painstakingly repeating these lists.

Thousands of such omens appear on the cuneiform tablets, and a number of them involve cats:

> *If a white cat is seen in a man's house—that land [will be seized by] hardship.*

> *If a black cat is seen in a man's house—that land will experience good fortune.*

> *If a red cat is seen in a man's house—that land will be rich.*

> *If a multicolored cat is seen in a man's house—that land will not prosper.*

> *If a yellow cat is seen in a man's house—that land will have a year of good fortune.*

The list of omens from Assyria seems to confirm the modern prejudice that folklore—and especially animistic folklore—is mere superstition. After all, cats of any color could no doubt be found in any house. But to apply the label *superstition* is to impose modern rules on people who thought in quite different terms. Ancient Mesopotamians did not, like us, expect all their facts to be empirical, able to be tested with the physical senses. Omens spoke the language of dreams and divinations; they tracked the speech of gods. It is not likely, in other words, that the omens were used in a literal sense to predict a literal future; they are not "superstitious" in the sense of making a simple irrational mistake regarding cause and effect.

Still, even if they aren't straightforwardly superstitious, they are striking in some ways. For one thing, they do assume connections between human beings and their surroundings, a holistic universe in which cats and land and people are all joined. Yet despite seeing connections, these omens pay remarkably little attention to the welfare of flesh-and-blood cats. The omen-cats have crossed over from furry to symbolic, leaving the real focus of the omens on human welfare. Cats are but vehicles for finding out about human business, a means for discovering whether human beings will prosper.

The scribes who copied these omens were of course preoccupied with prosperity, for they served the emperor, whose business it was to make the land flourish. However, the people who actually farmed the land and ensured the prosperity of the region were common people working the enormous estates of wealthy families, and the largest estates of all were those of the emperor. Through his nearly constant military campaigns he had dragged home booty from as far away as Egypt, including tens of thousands of people from distant lands relocated to Assyria and Babylon to become his slaves. Severe inequality thus existed in Assyrian society between a few ruling families and the masses of people who could not read and were likely owned as slaves or their labor owned in serfdom to the nobility.

Recall too that this land was physically kept in place through human control. It was a kingdom where good government traditionally meant digging irrigation canals and maintaining dikes and where people for centuries had celebrated the slaughter of Tiamat at every New Year.

I can't help but wonder if the anthropocentric quality of the omens has something to do with these practices of controlling lands and people. In a culture of dominance and control, whether of Earth or of human beings, does nature become merely a vehicle for human concerns, with animals valued less for themselves than for the information they can provide about human fortunes? In a culture where give-and-take is not part of the social equation, does religion too drift away from celebrating relationships of mutual benefit with the more-than-human world? I don't know enough about ancient Mesopotamian culture to begin to answer these questions, and the information may not be available more than twenty-five hundred years later.

What does seem clear is that the Mesopotamian culture of omens as well as the Mesopotamian pride in empire passed eventually to the whole Mediterranean region. Six hundred years later, in a Roman Empire founded on similar infrastructures of controlling water and earth, where a tiny minority of people regulated access to the gifts of the land, magical religion became the most popular expression of faith. Millions of people throughout the empire repeated omens and performed magical spells to manage all the out-of-control events in their lives, from unfaithful lovers to intractable illnesses to misfortunes of every hue. Their spells passed down through European history and became the tokens of animism that the elite were determined to wipe out at the dawn of the scientific revolution.

Yet those cats still survived, or at least the omens involving them. Though the culture of omens was suppressed in the 1600s, sayings about cats survive to our day, as if they had not merely nine but nine hundred lives.

༄

DAY BY DAY Brio worked his magic, digging himself ever deeper into my heart. His eyes, golden and astonished, gazed often in my direction. He settled on my lap now as often as on Tim's for his royal naps. In the yard he was as likely to curl up near me as to try to elude me. He had definitely chosen me as his own. But a human heart may take its time in opening, and though mine had definitely made room for Brio, I hadn't exactly fallen in love with him. Falling in love with cats was the business of goddesses—or at least it was in a story I once heard about ancient India.

In the king's court, it is said, a contest took place between a Brahmin and a Penitent. Which holy man was more powerful? The Brahmin, dressed in fine clothes—for he had been born to his sacred station and possessed all the powers of his caste—claimed that he was so powerful he could ascend to heaven, pluck a flower from the sacred heavenly tree, and bring it back for the king. Whereupon he disappeared from sight, only to reappear a few moments later. The members of the court gasped to see the precious flower in his hand.

The Penitent, itinerant and ragged, was the only one of all the people who remained unimpressed. He announced to king and court that he would show his power by sending his cat to heaven in his place to fetch the flower. Whereupon the cat began rising, rising, until he finally disappeared from sight.

The eager crowd waited, murmuring, to see if the cat would reappear with a flower. But in heaven something unexpected was taking place. When the cat arrived, the head goddess of all the forty-eight million goddesses of that particular heaven laid eyes on him and instantly fell in love with him, deciding in that moment to keep him for her own. Others reminded her what a grave offense it was to steal a cat, especially from a mortal. Yet she insisted, finally agreeing to a compromise: she would keep the cat for only three centuries.

At the king's court the people waited, but when the cat failed to reappear the gathered crowd grew impatient. The Penitent, knowing that the gods and goddesses were up to mischief—for he was a truly powerful holy man—placed a spell on all the members of the court so they neither hungered nor thirsted nor grew old nor died. And in this way they waited for three hundred years.

At the end of that time a great wonder was seen in the heavens. "The sky reddened, and the cat appeared on a throne in a cloud of a thousand hues." In his paws he carried not just one sacred flower but a whole branch from the heavenly tree. The Penitent, of course, was declared the more powerful holy man. And "the only incredible part of the story," concludes the storyteller, "is that the goddess should ever have permitted herself to be separated from a cat she had known and loved for three centuries."

I was beginning to understand her attachment. When Brio climbed onto my lap, I reveled in his lithe and silky frame. When the motor of his chest revved up in contentment, I felt more deeply at ease. And as my love for him grew, it began to take another form as well: now I too worried about Brio's bloody nose.

رؔ

WHEN PEOPLE LIVE by listening to the land instead of trying to remake their environment to suit themselves, what does their folklore sound like? As I am studying the ancient omens of Mesopotamia, an article crosses my desk about folk sayings repeated by contemporary Aymara and Quechua people in the Andes. These are the people who practice sweetening their relationships through nurturing and respect to ensure that they and the Earth they depend on can regenerate in perpetuity.

At first blush their sayings sound a lot like the Mesopotamian omens, many of them following the same "if-then" pattern of the ancients:

In a good year, the [Pleiades] stars shine brightly [on June 24, the first day of the agricultural year], but in a bad year they are misty or vague.

If the ground is dewy on . . . June 24, or if the smoke from the kitchen fires hangs near the earth, it will be a good agricultural year; . . . if it is dry that morning, or if the smoke disperses into the atmosphere, it will be a bad year.

If the fox howl is throaty, it will be a good year; if the fox howl sounds thin and high, it will be bad.

Do these sayings too belong to the obscure language of divination? Certainly the tone of a fox howl can't say much about the next growing season, can it? Anthropologist David Browman, who studied Andean omens, thinks otherwise. He explains that a fox howl sounds high and thin when the air is dry but low and throaty in humid air. A throaty howl—as well as dew on the ground and heavy smoke—signifies enough moisture in the air for a good crop year.

And what about stargazing? People of the Andes have been watching constellations since at least the time of the Incas to predict the next growing season, but can the stars say anything about the coming summer? A few years back anthropologists teamed up with meteorologists to investigate this question. They placed folklore next to satellite imagery, looking for possible connections, and found that high, thin cirrus clouds always increase over the tropics at midwinter (mid-June) of El Niño years but so thinly that they cannot be detected by the unaided eye. However, the added cloud moisture does slightly dim the sparkle of the Pleiades. El Niño is associated with drought conditions in the Andes, which means that Western and Andean observers agree: a misty Pleiades at midwinter foretells a dry and unproductive growing season. The centuries-old practice of stargazing still works.

Or at least it works when it is performed with the minute attention that Andeans bring to their relationships with others. In this culture, which regards nurturing and being nurtured as the bedrock of life, people try to pick up every signal, every speaking, from foxes and clouds, from stars and grasses and maize and llamas. They do also practice divination to discern the speaking of invisible deities, but in relation to the visible world, Andeans keep their thinking attached to bodies. Their beliefs about foxes reflect actual foxes; knowledge retains its link to the living other, and belief remains nested in relationship. Does their commitment to reciprocity and relationship keep their belief system from wandering away into the purely speculative and symbolic? Does this same commitment to relationship keep them practicing a social system with comparatively little inequality, where people look out for one another and the other creatures on the land? Certainly their commitment to close relationships with all others in their ecosystem has led to a traditional ecological knowledge that is as rigorous as the most precise technological instruments can measure.

So what about my own culture? If by *superstition* we mean irrational beliefs, where does superstition reside in the technologically advanced societies of the global North? I suggest we look at the very means by which modern people measure the well-being of their own societies.

Think of it this way: the ancient Mesopotamian omens as well as the Andean sayings about fox howls and stars were, or are, used as economic indicators, to help people gauge the potential prosperity of the region, whether the year will be a good one or a bad one. So how does my nation measure economic well-being? Through the gross domestic product (GDP).

The GDP is the sum of the economic transactions taking place during a specified period of time. As long as it goes up, people feel reassured, thinking the nation is becoming more robust and prosperity is increasing. The fortunes of political candidates and parties, not to mention a host of policy decisions, ride on its back. But does the number represented by the GDP have much to do with people's true well-being and happiness? As is increasingly being recognized, the answer is no. The GDP can rise even when the health of citizens or land is worsening. For example, the GDP rises when people get sick and require health care, for the total value of health-service transactions just increased. Similarly, the GDP can rise through deforesting land or removing mountaintops in mining or fracking for gas, because jobs as well as timber, coal, and gas are generated through those activities. Seeing the GDP go up, politicians and stock markets remain happy even though the foundation of an industry—the forest or the bed of fossil fuel—has just been destroyed and the health of land and water and local communities has been devastated.

Using GDP to measure national well-being comes uncomfortably close to using cats to predict the fortunes of a country. The GDP is about as cognizant of people's well-being as I imagine ancient cats were about the king's chances of increasing his wealth that year. In both cases, a theory is abstracted from the living world, a theory that slips ever further away from relationship with the reality it supposedly describes. The GDP ignores data about the destruction of nature, which leads to erroneous measures of human well-being. Given all the real-world information it disregards, can the GDP be considered a rational system of belief?

I wonder if superstition, or irrational belief, arises when humans forget their cousins the creatures, neglecting their relationships with salamanders and salmon, with clouds and rain, with soil and microbes and fungi—and cats. Certainly losing those relationships leads my society to believe it is progressing economically while it destroys the irreplaceable foundations of life. Similarly, is it the loss of relationship to other human beings, such as in situations of inequality, that prompts people to rely on beliefs that are contrary to reality? When people have little power over the decisions that affect their lives—about their neighborhood or the land where they live—do they turn to using nature in magical spells to control the overwhelming forces? Or, at the other end of the spectrum, when people of privilege are accustomed to controlling others, does their attention wander away from relationships with real animals, stars, or mountains?

Of the three economic indicators here, cats and starlight and GDP, the most rational one turns out to be the Andean system, for it is based on careful assessment of the natural surroundings as well as respectful relationships with other creatures. Only minute observation of nature and careful give-and-take with place can keep cultural beliefs residing close to actual bodies, and only beliefs that remain in relationship with living bodies can help us make good decisions about human—and Earth—welfare. Only when waters and whip-poor-wills and hickory and oak as well as human beings are respected for speaking and choosing can humans build a world that can be sustained through seven, and more, generations.

↜

IF FLOOD AND whirlwind are not choosy, venting their energies on anyone in their path, they are still in some sense choos*ing*. Weather researchers tell us that rain and clouds behave like predator and prey: the same graphs that describe relationships between the populations of, say, foxes and rabbits also describe the waxing and waning of clouds and rain. When clouds proliferate like rabbits, the rains, like predator foxes, also increase. The rains "eat" the clouds, diminishing them and beginning the cycle once again.

The conceptual leap in this case is modeling a weather system on animal behavior, and it is part of an increasing trend among many scientists to document self-organization, or the ability of organisms and systems to create pattern and order through thousands of small-scale, internal choices. Think of the complex precision of a honeybee hive or a school of fish in motion. No outside choreographer orchestrates the action; rather, order emerges through communications within the group. Each fish makes nanosecond choices depending on the placement of its partners and the direction the group is headed. Each bee contributes thousands of momentary choices, from locating pollen and communicating the directions to it through dance to collecting the pollen and nectar, storing it, and transforming it into honey. All together the bees create a stable and sustainable life, each of them responding moment to moment with honeybee decisions, a continual stream of individual choice and response creating the life of the hive.

Theoretical biologist Stuart A. Kauffman, best known for his writings on self-organizing behavior in nature, underscores the wonder inherent in the creativity and choice belonging to all living systems. "This web of life," he says, "the most complex system we know of in the universe, breaks no laws of physics, yet is partially lawless, ceaselessly creative. This creativity is stunning, awesome, and worthy of reverence."

Physicists say that the ability to choose and create extends to the subatomic level—that in the world of quantum mechanics a particle of matter (or energy) has to be prompted into "making up its mind," as physicist P. C. W. Davies put it. In the standard or Copenhagen interpretation of quantum mechanics, a subatomic particle does not have both a position and a momentum until the act of measuring it prods it toward having one property or the other.

Of course *choice* may not be the best word for an act that looks so little like human choosing. *Choice* implies consciousness, a noun heavy with Platonic notions of spiritual essences and static states of being, and we inevitably project human experiences of choosing whenever we use the word. But *choice* can also mean a simple act in response to immediate conditions,

like a fish darting in close harmony with his neighbors or a bee setting out
to follow her sister's dancing directions. The electron picks one state over
another—it chooses—when forced to do so by a tracking apparatus.

Following quantum theory to its logical conclusion strains credulity,
and for good reason. One of the originators of the Copenhagen interpretation, Niels Bohr, said that anyone who is not shocked by it does not
truly understand it. The implications are so wacky that even physicists
cannot agree on them.

This is how wacky. Suppose you look down a dark and foggy street.
The night and the murk hide the street; all you can see are small pools of
light in a row, each one circling a lamppost. An orange cat suddenly appears under one of the streetlights, rubbing her back against the base of its
pole. Then she disappears into the foggy night, only to appear moments
later in the glow of the next streetlight. Common sense tells us she walked
from one lamppost to the next, but in the quantum world things are not
so simple. Some interpretations of quantum theory claim that she (or her
quantum analog, a particle of matter or energy) traveled multiple, perhaps
infinite, paths to get from one destination to the other. Some say that when
she left the first pool of light she traveled simultaneously through many
universes, and only one version of her, the orange cat belonging to this
particular universe, showed up at the next streetlight. (I like to imagine
independent-minded cats enjoying both of these interpretations.)

But the Copenhagen interpretation, and especially Bohr's version of it,
was more radical than any of them. Bohr said, in effect, that it is wrong to
speak at all of the cat's path from streetlight to streetlight. We simply don't
know anything about her in between. We have no way of knowing where
she is outside of the lighted moments (an epistemological problem), but
more than that, outside of the light she may not be a cat at all (an ontological problem—and an interpretation that cats, if they care at all, must truly
enjoy). Like Macavity the Mystery Cat, she's just not there until the act of

registering her presence under the light brings her into being once again. Whether she is seen by a camera or a human doesn't matter; the point is, said Bohr, that the cat—okay, he didn't say a cat, he said "an elementary phenomenon"—*is not* a phenomenon until it is observed to be one.

In this universe, concluded Bohr, there is a radical entanglement of observer and observed. We are all observer-participants. The universe itself is participatory—a well-established principle in physics that the wider culture has yet to assimilate, though the idea is growing in some corners of science.

Maple trees and finches, bears and beavers: each of us is choosing, each responding, each creating. Together we bring one another and the world into being.

~

NEWTON'S CLOCKWORK UNIVERSE is gone, though we lumber on as if it still existed. Western cultures for the most part have yet to notice that the clock has been replaced by community.

A world of observer-participants is a radically communitarian one, not so different in some ways from the sacred hoop of the Lakota, where "life is a circle, and everything has its place in it," or the creation accounts of other indigenous peoples in which animals, trees, insects, rocks, plants, and humans together engage in a cosmic dance, generating the world moment to moment in a continuous stream of creativity. Even the physicist John Wheeler wrote, "It is extraordinarily difficult to state sharply and clearly where the community of observer-participants begins and where it ends."

In a participatory vision of nature, the old dichotomy between free will and determinism evaporates, for it depended on the idea of strict causality, either divine or mechanistic. Nature, it turns out, is not predictably causal, determined either by an omnipotent God or by mechanistic laws, but neither is it utterly random. There are more than two choices—as many as all the actors put together can generate. If the world is participatory, then every entity has agency—not in an absolute sense but in response to the suggestions of its neighbors, in the immediacy of each moment. We are all, like fish in a school, responding moment to moment to the choices of our neighbors, both human and more-than-human. And in a world where even the smallest creature, even a subatomic particle, is continually choosing, the results are bound to be unpredictable.

The future of the world, then, rests on a multitude of individual choices—those of pika and fern, human and sanderling alike. As writer and philosopher Susan Griffin observes, "It is possible to imagine a collaborative intelligence shaping form, event, circumstance, consequence, life." By entering this network of creation and response, taking our place as cocreators, human beings might remember what it is to feel at home in the world. "One is no longer placed in an alien environment," Griffin continues. "Instead, in and through existence one enters community."

⌒

As I MADE plans in summer 2006 to relocate, my brother in Ohio was also undergoing deep changes. In recent years he had taken steps to clean up his decades-old anger at Mom. I remembered his phone call the previous year—he a truck driver now, rolling one evening through rural Michigan, me sitting at my desk in Grass Valley. "Well, Sis, you're going to be happy to hear this," he said from his cab. "The counselor I've been seeing—she's been pushing me to make peace with Mom. So I went out last week, late one night. I went out to where no one could hear me, and I just beat the ground and bawled and bawled. I forgave her everything."

I was glad to hear it, though I doubted that one night of weeping would clear out a lifetime of rancor. But in the months since then, there had been a new quietness about him. Still jovial, still telling his favorite dirty jokes, he had seemed a little less blustery. More peaceful. He no longer insisted in anger on being top dog in his house. He tried to show his wife how much he loved her, though he didn't fulfill her wish that he stop smoking. But he did grow softer, more accepting, as if he too was beginning to glimpse a deeper communion of all, as if his life was rooted less in some rigid, ledger-checking God and more in the daily loves of life—in being kind, in providing a listening ear for his own or his wife's children, in gently rocking his toddler grandchildren.

Then Bro went into chemo. One night he had to be rescued from his semi by ambulance and taken to the emergency room for disabling abdominal cramps. Tests showed colon cancer. Still he seemed indestructible, missing little time at work, barely allowing the chemo to faze him. While I was readying my house for sale, Bro was mowing his enormous lawn carrying his chemo drip over his shoulder. After treatment he was given a clean bill of health, and he went on with his life. Nothing more was said about cancer.

Later we visited my brother and his wife in Ohio. Bro seemed a little more sober than I remembered. There was a deeper kindness. When someone during the visit, probably me, mentioned another person in a disparaging tone, he interrupted in an offhand way: "We probably don't need to say that kind of thing about anyone." He was eager to share two songs he'd written while rolling down the highway, the latest in a line of worship songs he'd composed in the last few years and taught to his local congregation. He picked up his guitar—never far out of reach—and sang: "In my Father's house, there's room for everyone, in my Father's house."

The scales of my brother's life had tipped at last fully toward love.

⌇

WITHIN MY HOME too, gentleness was growing. Brio was beginning to slow down, his nosebleeds, as it turned out, caused by nasal cancer, an ailment of older male cats. But as he slowed, day by day he grew sweeter.

The indignant, demanding sides of his personality fell away as if outgrown and now shed. He turned patient and peaceful, becoming by far the most resilient one in the family. When we sat on the sofa, he nestled happily in our laps, stretching out one soft warm paw to touch an arm or rest it on a human chest. When Tim and I sat together, he draped himself across both of us at the same time.

Brio and I continued exploring the outdoors together. When I wrote at the patio table, Brio occupied my lap. If I lounged in grass reading a book and Brio was let out the back door, he made a beeline toward me, settling happily a foot or two away. One day when he was enjoying the yard on his own, a doe ambled in. They walked carefully toward each other, sniffing, and nearly touched noses before she turned away.

To skip ahead a bit in our story, in his final year of life Brio would get to know a new puppy, who joined our family after we moved to Boulder. Bodhi was an energetic blue heeler mix ready to herd whoever happened to be nearby, and now and then he would succeed in nudging even the elder Brio into a good-natured chase around the house. With the puppy Brio would be at first playful and then, over time, calmly patient, and finally he would find fulfillment in sitting peacefully on the sofa for hours next to the newcomer. Our friend Mimi summed it up best: "The sweet hellion has turned into a sweetheart."

With the new puppy, Brio would rediscover the joy of grooming a friend. When Bodhi, three times Brio's size, approached him on the sofa and stuck his head under Brio's nose, Brio instantly reached out his tongue to lick the puppy's face—first the left cheek, then the right, then the ears.

In Brio, it seemed, there was nothing left but love.

1 2

Earth

When we see land as a community to which we belong, we may
begin to use it with love and respect.

—ALDO LEOPOLD,
FOREWORD TO *A SAND COUNTY ALMANAC*

"HEALING COMES FROM the land." The Maidu elder's voice was
firm yet quiet.

It was a sunny day in October, and I was driving home from downtown Grass Valley, my car radio tuned to the Indigenous People's Day
celebration taking place across town. I had vacated the house for a couple
of hours so that yet another prospective buyer could walk through it. We
were impatient to sell and be on our way; the cottage had been listed for
three months already, and winter was coming on. This longer-than-usual
sale time was no surprise because the housing market had begun to cool,
but in the fall of 2006 few people could see just how dramatic that down-
turn would become.

The Maidu man repeated for emphasis: "The earth is the source of all
healing."

My mind drifted back to the time when I learned that to be true in a literal way. It was the time of grieving for my parents and my marriage and what seemed like a hundred other losses. An artist friend, seeing my depression, introduced me to mud: she led me to her studio and showed me how to build things with clay. We rolled the clay into a smooth slab, measured and cut five pieces for a simple box, scored their edges with a fork, and glued them together with the silky watered-down clay called slip. When my box emerged from the kiln intact, I was hooked. Earth had wedged itself under my fingernails and under my depression as well. I found an inexpensive ceramics class and signed up.

During those three hours each week I could forget about words and think instead with skin and fingers and earth. Over months and then years of classes, I watched my mind quiet while my hands played in clay.

Early on in the class, I decided to fashion an espresso cup and saucer for Mimi, to add to her collection. Could I shape it like a rose—the saucer a sloping leaf, the cup an opening bud? My hands worked and worked, but nothing looked remotely like the flower.

Absorbed in the task, I didn't notice that half an hour had passed before I finally discovered the rosebud—a gentle fullness in the bottom of the cup rising to a softly curving edge. Now, how to shape a few petals near the top and still have a lip on the cup for drinking? I tried one design and then another. After going round and round the tiny cup, my fingers at last found the hint of petals near the rim.

Suddenly I realized how hard I was concentrating and looked up. A woman across the table was making a small jewelry box, carving circles and dots into its sides. Another was painting yellow and black fish fins onto her freshly hollowed-out oval platter. All the usual scenes of the clay class.

Yet something was different. I sniffed the air. A hint of rose essence had appeared at my end of the table. I thought I was imagining it. I said aloud, "I just got a whiff of a rose." The woman next to me said, "Yes! I've been smelling it too!" We laughed, delighted as children. Earth gladdens; earth is magic.

ᕐ

DEBORAH BIRD ROSE is an anthropology professor in Australia who spent years studying with the Aboriginal people in the settlement of Yarralin in the Northern Territory. She says the Yarralin people view land as a "nourishing" place. Country—including sky and sea and ancestral creators and underground as well as surface ground, with all its plants and animals and waters—is a place of relationship. *Country* is a proper noun. "People talk about country in the same way that they would talk about a person: they speak to country, sing to country, visit country, worry about country, feel sorry for country, and long for country." And country responds in kind. "People say that country knows, hears, smells, takes notice, takes care, is sorry or happy."

Rose stresses that this use of *country* is not like "going to the country" or "spending a day in the country." Instead, country is living; country has will and intention, which means country has both a past and a future. It is not inert but full of history, full of meaning. "Because of this richness, country is home, and peace; nourishment for body, mind, and spirit; heart's ease."

Prior to the coming of Europeans, every square inch of land was known to the people and loved by them. Each watering hole, each hiding place of game, each creek and anthill had story. Humans actively managed all of it, primarily through fire. People performed controlled burns in cycles of years, burning small patches of land to create a finely detailed mosaic pattern across the country. Fire purified the land and helped to keep the whole in balance. It functioned as a form of agriculture, encouraging the edible plants and discouraging the trail-choking plants. Burning was also a hunting strategy, removing overgrowth so that animals could be more easily seen, as well as a conservation strategy in that it encouraged habitat growth. Australian animals who have gone extinct or are on the verge of it are endangered in large part because of the loss of Aboriginal burning.

Because land was burned regularly and the burning timed over cycles of years, the fires were of low intensity. "Big fires come when that country is sick from no one looking after with proper burning," explained one Aboriginal person. Country that had been carefully burned was well-cared-for country. And this is key to their management: they aimed to promote the welfare of the land. Humans would benefit, yes, *but not humans alone.* The advantages of human tending would flow toward the more-than-human inhabitants as well. It had to, for of course all were sharing the same home.

Aboriginal management practices give the lie to the old conservation idea in the West that land, to be healthy, must be untouched by humans. No surprise, then, that Aboriginal definitions of wilderness run counter to Western ones. Deborah Bird Rose tells about traveling with Daly Pulkara, a Yarralin man, to a place where overgrazing of cattle had caused great heaps of soil to erode down a hillside. Rose asked Pulkara what he called this land. "It's the wild. Just the wild," he said heavily.

The land had been made "wild" not by letting it be but by trampling it, not by the absence of humans but by their irresponsible presence. To Pulkara, "wild" land was land that would wash away in the next rain. "Quiet" land, by contrast, was land well cared for, loved and nurtured throughout the generations.

Intimacy is the foundation of Aboriginal relationship to country— intimacy with the finely tuned details of a healthy ecosystem. But more than an ecosystem, country is kin. Intimacy with land means interdependence with all the relatives who share the same territory—the animals, plants, waters, and hills, the clouds and ancestors and original creators, who are all present, all communicating. As in a family, the fates of all are intertwined. Humans are obligated to nurture land so they can be nurtured in return. "If you don't look after country, country won't look after you," said April Bright, a Yarralin woman. When land is cared for, relations between people and country are friendly; when humans are irresponsible, taking more than their share so that animals or plants or waters suffer, the bonds of reciprocity are broken.

The law they live by, say the Aboriginal people, resides in the land itself; it was not made by humans. In this way it is different from the law of the Europeans. "Whitefellow law goes this way, that way, all the time changing," said Doug Campbell, a senior Yarralin man. "Blackfellow Law different. It never changes. Blackfellow Law hard—like a stone, like that hill.

"The Law is in the ground."

⌒

IT IS TEMPTING to think the Aboriginal approach to country represents an earlier strand of human development. Didn't all peoples in some dim and forgotten past live intimately with land? Popular ideas about human prehistory suggest that hunter-gatherers killed in balance with their surroundings; they listened for the whirring of bird wings and worshipped the life-giving waters; they gathered the medicine plants of their locale. Contemporary indigenous peoples, in this view, have simply managed to hold on to a way of life that modern society has lost. Indigenous cultures can teach modern people what it was like in the time before people were separated from nature.

I do believe in learning from indigenous cultures, but not because they provide windows into an earlier way of life. To see them as representative of an earlier era is but another face of the colonial arrogance that sees industrial societies as more advanced. Indigenous societies too have undergone thousands of years of cultural development. What they deserve credit for is the hard cultural work they perform over and over again, with every new generation, to arrive at a way of life in balance with their surroundings.

What if living in harmony with Earth is always an achievement? What if it always involves choice? If it is true in a human family, it must be true with the more-than-human relatives as well: living well together takes work—also humor, affection, and careful attention. Falling out of harmony with one another happens easily, as easily as forgetting.

Human beings are not "fallen," in the sense of being unable to choose what is right. What we are is forgetful. We forget that our fate is intertwined with that of others, our good life dependent on the good life of myriad others. We forget the body, placing our mental constructs ahead of the needs of physical well-being, allowing ideas to drift away from relationship with realities—and creatures—on the ground. Without continual reminders, we tend to forget the connections that bind all together in a single interwoven web.

Living sustainably is learned behavior. In this sense it is indeed ethics, as pioneer conservationist Aldo Leopold wrote—not because "ethical" behavior is different from "instinctive" or "natural" behavior but just the opposite: because living in relationships of stability with all the surrounding creatures is the most natural thing in the world to do. It is how every other species lives without thinking about it. We are the ones who forget. We may be the only ones who *can* forget. We clearly are the only ones who can override—through our technology and values—the delicate balance that keeps the land-community operating in highly organized circles of interdependence. And we are certainly the only ones who legislate into our collective life such a narrow pursuit of self-interest that we imperil the healthy functioning of the whole community.

For in this too Leopold was spot-on: the Earth to which we are born is a land-community. And this community is threatened now, as it was in 1948 when Leopold wrote, by defining the good life as the pursuit of economic self-interest. "We abuse land," he said, "because we regard it as a commodity belonging to us." Laws that favor self-interest instead of balancing self-interest with responsibility and obligation toward others lead to massive ecological destruction. "When we see land as a community to which we belong, we may begin to use it with love and respect."

~

WHEN DID HUMANS begin to forget their cousins the creatures? When did we fail to remember that the web of life is a delicate one, requiring attention and care? Some point to the rise of agriculture ten thousand years ago. Ecologist Paul Shepard suggested that domesticating plants and animals led us to turn "from finding to making," from taking our chances with nature to manipulating nature. Others say that when people gathered into cities and built urban centers we became increasingly separated from the natural world. Environmental historian J. Donald Hughes writes that the urban revolution meant "the great divorce of culture and nature" wherever it took place on the planet. Still others say that literacy trained people away from intimate connections with the more-than-human world. Philosopher Eric Havelock observed that when people no longer had to "story" their experiences, as they do in oral societies, telling tales of characters and relationships, they shifted to considering others as things rather than persons. Cultural ecologist David Abram emphasizes that relying on the printed word changes our ways of perceiving: instead of listening to breezes, watching clouds, or feeling our way along animal tracks—all practices to cultivate intimacy—we allow our senses to dim, except for one particular way of using our eyes.

While there is truth in all of these analyses, I want to point to something at once simpler and more sweeping. I think we forget our cousins the creatures when we forget each other. When we retreat from caring for the human community, we lose regard for the more-than-human one as well. And the opposite is just as true: when we fall out of relationship with the natural world, we lose interest in helping one another thrive.

For this is the bottom line of survival: it depends on our relationships with others. Though the land-community survived for millions of years without humans, we cannot survive without the land-community. We are dependent for our day-to-day survival, our very existence, on billions of nonhuman others. And we are dependent in equally complex ways on one another. This is the reality that sustainable societies, and especially indigenous cultures, train themselves and their children to remember.

Anthropologists tell us that hunter-gatherers typically share resources more equally than other types of societies. One recent study of several contemporary hunter-gatherer groups showed that they tolerate only as much inequality in income and wealth as in Denmark, currently the most egalitarian country in the world. Is it the face-to-face quality of life that keeps them remembering interdependence? Not likely. Sometimes face-to-face contact can make equality more difficult: can you imagine urging the directors of your company or university to bring their incomes into line with everyone else's? Is it the solidarity that grows when people must help each other survive in a precarious natural world? Is it their nomadic lifestyle, which prevents them from getting too attached to possessions? The fact that Scandinavian societies come close to hunter-gatherer levels of equality would indicate no. The Yarralin people notice that white folks are uncommonly serious about their possessions; the Europeans, they say, can't take a joke about their property.

Hunter-gatherers practice remembering what most Western cultures have forgotten: that the deep wisdom of Earth flows in cycles and in cycles of cycles. Life in the land-community means engaging in give-and-take.

ᕒ

I STAND NEXT to a cliff overlooking the ocean. The day is sparkling, warmth radiating from nearby rocks though my jacket is zipped to the chin against a stiff and chilly breeze. A line of pelicans flows low across the water, only a wingbeat away from the waves, taking advantage of the cushion of air that bears them along just above the surface of the sea.

The sky is so blue it hurts. Darkly joyous waters crash again and again into black rocks just offshore. I set out walking along the shore into the wind. My shoes sink into the sand; plowing forward takes effort. Thoughts trail into reverie as I slip into the rhythm of left foot and right foot.

Every wave that flows also ebbs. The water that falls from the sky tumbles and percolates to the sea and is taken up again to sky. Molecules collect for a time into one form and then scatter, only to collect in a different form. Food becomes fertilizer and once again food. One species proliferates, only to be checked by the rise of another. Every striving in summer is followed by resting in winter, every daytime followed by night. Each gathering into birth also disperses into dying.

Taking breath in, giving it out. Left foot, right foot.

⌒

THERE ARE AS many ways to practice give-and-take as there are human societies to imagine it, and almost all of those ways are intended to help people remember that they are not alone—that they are part of a much larger, breathing land-organism. Caring for land by rotational burning, as Aboriginal peoples do, is but one example. Caring for land, to them, also means caring for the people who live on that land. They believe that owning land includes the obligation to share the land's resources. In Aboriginal country, people are free to enter a territory that is occupied or owned by others, but as guests they must ask permission to take of its resources. The owners of the territory, as hosts, are required to share with those who ask. Ownership does not entitle one to an added bonus from others, as it does in a Western economy (in the form of rent or interest); rather, ownership binds one to share with others. "To own is to have the obligation to share," as one observer put it.

If we live on a finite planet, sharing makes good economic sense. When the pie is limited, sharing it is the only way to make sure everyone gets a piece. "Distribution is not just a moral issue, it relates to efficiency," says ecological economist Joshua Farley of the University of Vermont. As a grad student at Harvard and Cornell, Farley found himself asking questions that none of his professors had answers for, questions such as, If physics tells us that everything in nature inevitably winds down (the second law of thermodynamics), why do we think an economy can experience unlimited growth? And why does conventional economics treat all of Earth's resources as if they were replaceable? "You mean, when we run out of clean water, we'll just find something else to drink?" he asks.

If we are going to live within the planet's limits, Farley says, then we have to come to grips with how its gifts are distributed so that all have what they need. Vast inequalities in wealth are costly. For example, societies with poor distribution of wealth also suffer poorer health, for not only does poverty increase stress, leading to more health problems, but *relative* poverty does so as well. Lower-status people have poorer health even when they are well enough off to meet their basic physical needs.

The problem in our present economic system, he says, is that a massive redistribution of Earth's gifts is indeed going on, but it's happening in one direction only. The goods of Earth, such as access to clean water, to forests, to energy sources and airwaves, are being funneled toward fewer and fewer people. The challenge is to allow these benefits to move in the other direction as well—or in economic terms, to recapture the income that accrues to the few when they use the benefits of Earth, which belong to all.

I hear him to be saying that we need to promote giving as well as taking. To live by the law of the ground, we have to share.

⌒

WHILE I AM writing this chapter, the Indian philosopher and policy analyst Daniel Wildcat, a Yuchi member of the Muscogee Nation of Oklahoma, comes to town. He is kindly, avuncular, but with a sharply focused message. "You've heard about the ATM model of family, right?" he asks. "You know, the relatives who only show up when they need something, to withdraw funds?

"Well, that's the kind of relationship we've been practicing with the Earth. We've been making a lot of withdrawals in the last two hundred years and haven't put much back."

Modern society, he says, needs to learn from the First People, who saw air, land, and water as relatives.

"You don't treat your relatives like resources."

ᔊ

FIRST NATIONS PEOPLE of the Northwest Coast practiced give-and-take through the potlatch, a system of laws, dances, feasts, and giveaways that was in place for two thousand years. Ronald Trosper, an enrolled member of the Flathead Reservation and a Harvard-trained economist, argues that the potlatch system ensured an abundant supply of salmon for two millennia—an astonishing accomplishment in sustainability. When Europeans arrived they found prosperous communities of healthy people living in substantial wooden houses and enjoying a culture that was rich in both art and trade. What the white people failed to realize was that the coastal natives "had actually manipulated their environment to its high level of productivity." The coastal people had practiced a particular kind of caring for their environment—for the roots and berries and fish who provided their food.

That caring consisted of sharing wealth with each other in public, prescribed ways. The coastal people organized themselves into "houses" consisting of several families, each house headed by a "chief." Each house owned, or had exclusive access to, a certain territory, which might include fishing sites, berry patches, gardens, and trees. As among the Aboriginal people, owning a territory mandated sharing its gifts. The coastal peoples shared by hosting periodic feasts and giveaways. A house would accumulate wealth for several years then invite other houses and villages to celebrate a marriage or a birth or a death, providing all the food for the feast as well as gifts such as blankets for all the guests.

Because the feasts were public affairs and all the gifts were publicly counted and announced, it soon became obvious if a particular house was not doing a good job of being generous. To keep their positions, chiefs had to meet the communal standard of generosity. And in order to have enough surplus to share lavishly, they had to take scrupulous care of their food territories. Each house was obligated to the salmon to consume the salmon's gifts sparingly and share the salmon's gifts generously, or the salmon would fail to return in the future.

The society, in other words, set up incentives for generosity rather than selfishness. Because all were expected to share, all took part in stewarding their resources—or as the coastal people prefer to say, they took part in caring for their relatives the salmon. Rewarding generosity rather than self-interest then encouraged more generosity—not primarily because it felt good but because it was embedded in the most basic practices of exchange.

Best of all, in such a system of "symmetric generosity," something magical happens. When people are rewarded for being generous, they tend to take from their environment only the optimal amount—the amount that balances with what others in their circle of exchange also need. Each house in the Northwest Coast economy, without consulting with others ahead of time, made environmentally appropriate choices. Each caught and consumed the amount of salmon that best served the broader community, including both the human and the salmon relatives (or economic partners). Of course their behavior is not magical at all; wise choices resulted from the incentive to share instead of hoard, an economic incentive that Trosper describes with mathematical models.

The potlatch system could be sustained over such a long period of time because it matched a fundamental reality of the land-community: that living on Earth means being bound up in a richly interdependent and complex web of creating and eating, giving and taking, sharing and collecting. By replicating interdependence in their human relationships, the coastal people preserved their unity with land. They created no separation between their ways of operating and those of the land.

Trosper suggests that some features of the potlatch system could be translated into the modern world. In a modern potlatch system "houses" might become "firms," and "chiefs" an "ecological review board" made up of representatives of each firm. Within a given ecosystem, such as a watershed, each firm would work to build up wealth, as corporations do in our present system. However, they would be entitled not to their full profits but to only a portion of them. At the same time, they would be entitled to a portion of the profits of other firms in their watershed. In effect, each firm would own part of the wealth of the others.

To implement such a system would mean moving toward the belief that property brings responsibility rather than privilege. It would mean redefining ownership as trusteeship rather than absolute control. It would mean, as a society, deciding to reward profit sharing rather than endless private accumulation. To say that this would require massive shifts in many fundamental beliefs and practices of the modern world is an understatement.

But a system of reciprocity also provides a glimpse into a possible third way to do business—something different from both markets and central planning. Today these are the only two economic models ever placed on the table, with the proponents of each decrying the other system's failures. Lovers of the free market point, with reason, to the authoritarianism of centrally planned systems. They complain that central governments only hamper individual freedoms. Those who support a stronger hand of the government in the economy counter, with reason, that free markets are ruthlessly competitive, pitting individuals against one another in a zero-sum game that funnels profits upward toward a privileged few.

What few people mention is that both systems stem from the same history. Marxism and capitalism alike are rooted in the modern industrial story of atomism—the isolation of humans from one another and from the rest of the land-community. Both systems treat the members of the land-community as resources rather than relatives because they share the modern story of a passive and inert Earth. Both set the interests of individuals and communities in opposition to each other, for both are founded on the story of human beings as selfish at their core. And both systems tend to stifle individual creativity, requiring people to abandon their own internal rhythms in favor of some external order. Centrally planned systems do it through imposing uniformity from above, while free markets transform citizens into consumers, squeezing individual preferences into the mold of mass culture.

In a world linked by ozone depletion and climate change and seas that are sick, nothing could be more urgent than healing that tired old split between individual and community. Our survival today is threatened most by imagining the welfare of the individual as separate from that of others.

A system of reciprocal generosity represents a third way lying outside this history of opposition. It joins the interests of individuals and groups and is rooted in a story of connection and interdependence. It honors the needs of all by honoring the ecological rhythm of give-and-take—the practice of sharing as well as receiving. In a reciprocal system, when you share with others, you can depend on others to share with you.

༄

MY CULTURE TOO has a heritage of sharing and redistribution, though it has been forgotten for thousands of years. The Bible I was brought up to read daily tells how the ancient Israelites instituted a special year out of every forty-nine, a sabbatical of sabbaticals, to restore equilibrium in their community. During a regular sabbatical, every seventh year, fields were to lie fallow so the land could rejuvenate. But during a Year of Jubilee fields that had changed hands were to be returned to their original owners, all debts forgiven, and all slaves and servants set free. No one would go destitute for want of land, no one starve while others luxuriated. Resources would again flow to those from whom they had previously ebbed, and the cycle would be complete.

Renowned economist and Mesopotamian scholar Michael Hudson says that the Israelite Year of Jubilee was borrowed from Assyrian and Babylonian kings, who from time to time announced "clean-slate laws" to prevent the unrest and migrations that take place when people are forced off the land because of debt. The decrees had conservative intentions, in other words, not utopian ones; it was in the state's interest to promote social stability—to keep the militia well fed and the farmers on the land. By the time the Israelites instituted clean-slate laws, after they returned from their Babylonian exile in the late sixth century BCE, the Mesopotamian tradition had been going strong for two thousand years already, begun by the Sumerians of about 2400 BCE, who called their laws *amargi*, literally, "return to the mother." In most cases the debts being canceled were those owed to the state, so in forgiving its own citizens, the state was restoring individuals to economic self-sufficiency—a stabilizing measure. Often debt forgiveness was the first act of a new ruler, to consolidate his popularity and power.

What the Israelites did was to give these decrees a different meaning. Instead of seeing the state as the ultimate landowner, they regarded the land as having a more-than-human origin: God alone was the landowner. "The land must never be sold irreversibly because the land belongs to me," God says in Leviticus. "You are only guests and tenant farmers working for me" (25:23). God tells the people to remember that they only borrow the land, and therefore they must use it wisely, making sure that everyone in the community retains the ability to support themselves. The other Israelite innovation was to set up debt forgiveness as a regular occurrence, taking place automatically every five decades rather than at the whim of a ruler who wished to curry favor.

Observing the Year of Jubilee weakened over time, says Hudson, as land consolidated in the hands of fewer and fewer owners. By the first century of the common era fields and vineyards in Palestine, as throughout the rest of the Roman Empire, were owned by wealthy absentee landlords—state officials, members of an aristocratic set of families cozy with Rome—who leased it to peasants to gain income from what the land produced. Peasants who fell behind in their produce payments and went into debt were driven off the land, which was then devoted to cash crops, such as dates and wine grapes and olives, instead of being planted with staple crops to feed local residents. Once the land belonged to only a few, the idea of returning it to its original owners in a Year of Jubilee changed into an out-of-reach utopian ideal.

⌒

BY THE TIME Christianity became the religion of the Roman Empire in the late 300s, the system of rule by patrician families was well entrenched. Abusive lending practices and land grabs had become the norm in an empire that provided no mechanisms for sharing profits, not even a property tax (though the wealthy were expected to engage in patronage). During the last quarter of the fourth century, as the empire was splintering, the process of consolidating wealth in private hands accelerated. Less than 5 percent of the people now owned more than 80 percent of the empire's lands and made most of the governing decisions for the empire. The system of holding vast wealth in the hands of only a few families bled the empire dry of funds and paved the way for its fall a few decades later. It also set the course for the more than a thousand years of serfdom to follow, for the enormous private landholdings became the founding estates of feudal Europe.

Ambrose, bishop of Milan, the capital of the empire, came from one of those wealthy landowning families. Given an elite education in law and classical literature, Ambrose was governor of Milan when, in 374, he was drafted by public acclaim to become bishop. He had no training in theology, was not even baptized, and he resisted the call with every fiber of his being. To make a public show of how unfit he was for the job, he returned to his judge's chambers and resumed handing out torture, the usual Roman sentence in criminal cases, and then invited prostitutes into his home. Was he struggling with the renunciation that he felt would have to take place to become a spiritual leader? His resistance was futile; the emperor himself ordered Ambrose to the post of bishop, and in the space of one week Ambrose was baptized and consecrated to each of the successive levels leading to bishop. Soon after, he donated his share of the family fortune to the Church, distributing some of it to the poor. Now that Christianity was endorsed by the emperor, more and more of the politically powerful, landowning bureaucrats were converting, but Ambrose was one of the first to give away his property when he became Christian.

A generation older than Augustine, Ambrose had been bishop for a decade already when the twentysomething Augustine arrived in Milan in the 380s to begin his career. A sober-looking man with dark hair and soulful eyes, Ambrose was a fervent preacher known for his flowing rhetoric in an age that measured social status and moral refinement by public eloquence. It was the sermons of Ambrose that so inspired Augustine that he at last converted to Christianity.

In many of his sermons Ambrose targeted the growing chasm between rich and poor. Today few people are aware that early Christians worried about wealth and poverty. Even fewer know that before Christianity became the religion of empire, which means before the Church itself owned

vast tracts of land, many Christians, including Ambrose, the emperor's
bishop, opposed the gathering of so much property into so few hands.
(After Christianity won state support, many wealthy converts donated
their estates to the Church so that it soon mirrored the crumbling empire's
immense gap between rich and poor.)

In one especially passionate sermon delivered probably in the late
380s, perhaps one of the very sermons that inspired Augustine, Ambrose
took as his text the Old Testament story of Naboth and his vineyard (1
Kings 21). A poor man named Naboth owned a small vineyard that, un-
lucky for him, sat next to the king's palace. One day the king decided he
wanted the vineyard for an herb garden and offered to trade Naboth for a
different vineyard. Naboth was hesitant to part with his family's heritage
and reluctantly said no. But it is dangerous to refuse a king. The king took
to his bed and pouted, refusing to eat, and the queen, seeing his unhappi-
ness, schemed to have Naboth executed.

Ambrose seized on the story and made it contemporary. "How far, O
rich, will you pursue your insane greed?" he railed. He quoted the prophet
Isaiah, who preached woe to those "who add house to house and join
field to field" so that they are the only ones living on the land (Isaiah 5:8
NIV). "The land was made to be shared among all, rich and poor alike," he
declared. "Why then, O rich, do you claim exclusive rights to the soil?" It
was an offense against nature to amass property at the expense of others.
"The world was created for all, but you rich men are trying to reserve it
for yourselves. And not just the earth but the sky itself, the air, the sea—all
reserved for the use of the rich few.

"Greed is whetted, not quenched, by gain!" he thundered on.

The remedy, said Ambrose, is to distribute the goods of the Earth.
"Sell gold and buy salvation," he urged. Share the proceeds with others.
Remember that God is the ultimate owner of all goods. Follow the words
of Jesus himself, who advised the rich young ruler: "Sell what you have

and give to the poor, and you will have treasure in heaven" (Luke 18:22). Ambrose ended his sermon: "If you want to be rich, become poor in the world, and you will become rich in God."

Ambrose was speaking from experience; he had been that rich young ruler, one of the few who followed Jesus's instructions. For the former judge and governor, becoming a Christian bishop meant from that point on teaching people to share the goods of the Earth.

୰

IN AMBROSE'S TIME, giving away property made a dramatic statement— dramatic because all the rules of commerce rewarded the opposite strategy of accumulating as much as possible for oneself and one's children. The functioning of the Roman world, from finding one's daily food to running the towns and cities of the empire, depended on following those rules. And so the preaching of Ambrose had little impact. Exhorting the elite to give away their wealth made no headway at all in a society where incentives were weighted in the opposite direction.

The sermon of Ambrose marks a turning point in the history of the Church. After the time of Ambrose, Christianity, which had just become the official religion of the empire, grew quickly into its new role as the unifying force of a crumbling regime. Transformations in its outlook and theology took place in the space of a generation or two. Greed came to be imagined more and more as an internal spiritual flaw, a case of what Augustine, only a few decades after Ambrose, would call concupiscence, or "lusting after"—the hallmark of original sin. As soon as greed became an individual spiritual failing, generosity turned into a remedy for sin. Giving away riches was transformed from a public act that challenged social and political inequality, as it still was in Ambrose's day, to a private act of personal piety.

The Church was not quite there yet in the time of Ambrose; in his sermon Ambrose instructed the wealthy to share the goods of the Earth to improve the lot of the poor and move society toward greater equality. But in the decades to come, much of this vision for equality was diluted as Christianity adopted the powers of empire and then became a feudal landowner in its own right.

The problems of a politically splintering, economically broke empire were not going to be solved by a few more wealthy bureaucrats giving away their property to the poor. The changes needed went far deeper than that. The empire needed nothing less than to address the rules that had enshrined such vast inequality in the first place, the same rules that had led to exploiting the lands of the empire beyond what they could sustain.

⌒

To DWELL IN relationship with a living Earth means to recast greed once again. More than a spiritual flaw within an individual person, greed is a failure of relationship. It is a symptom of forgetting—failing to remember that there are others, that one is not alone. It is a lack of reciprocity, an assumption that the benefits of the Earth can flow continuously in one direction without being given back or cycled again into the whole.

It is treating the Earth as resource instead of relative.

⌒

ON THE CAR radio the Maidu elder was giving thanks to the various parts of creation:

"We give thanks to Earth, for she supports us always.

"We give thanks to the grasses, for being soft under our feet.

"We give thanks to the trees, for keeping us warm in the winter and for sheltering us in our houses.

"We give thanks to the birds, for their sweet songs."

I was lulled into gratitude without thinking about it. To the immense cedar and pine trees lining the road—thank you for purifying the

air, especially along this highway thick with diesel fumes. To the soil—
thank you for nourishing all these trees, for keeping these woods lush with
greens. To the sun—thank you for the heat and light that make our lives
possible. My thoughts quieted, my mind settling into a more peaceful state
than usual.

Gratitude is one time-honored way to practice reciprocity. But much of
the time gratitude toward nature gets funneled, not toward Earth, but toward a deity residing apart from Earth. I think of the photograph splashed
across my computer desktop, taken in southern California in the spring of
2005, after a winter of torrential rains. The hills west of Bakersfield are
spread with wildflowers of every hue—a carpet of lavender lupine, golden
fiddleheads, and scatterings of white and pink on the left-hand slope; on
the right, a dense mat of lemon yellow broken with patches of more purple
lupine. The photo is stunning: green grass completely obliterated by a rain-
bow of colors—more wildflowers than I have ever seen in one place.

The photo made the rounds on the web recently, accompanied by the
caption "When God spilled the paint." The image of the caption is equally
arresting: a heavenly person standing over the ground splashing color from
above—as if the land is but a passive receiver, lacking the intelligence or
creative will to produce such a scene of splendor.

And gratitude toward Earth makes no appearance at all in the calcu-
lations that guide our commerce—as if the Earth is but a resource to be
used up at our whim. "Sometimes in June," wrote Aldo Leopold, "I see
unearned dividends of dew hung on every lupine. . . . Do economists know
about lupine?"

⌐

OUR RELIGION AND economics alike have failed us. They have colluded
in leading us away from this world, telling us stories about ourselves and
Earth that undermine our efforts to connect with one another and with

the more-than-human world. Believing that human nature is selfish and competitive and that matter is dead and deficient, the modern world justifies economic practices thousands of years old—of raiding Earth for material goods and using those goods to separate people from one another.

Today fantasies of heaven, in religion, accompany fantasies of an inert and passive Earth, in economic life. It is no accident that these fantasies flourish side by side. They are two faces of the same coin—a belief that abundant life can be found only apart from Earth. The belief arises more easily in settings where a comparative few enjoy the plenty of Earth. When severe inequality separates people from one another, people separate more easily from the more-than-human community as well. Poverty is incompatible with sustainability. So is great wealth.

Different choices are possible; different stories can be told. It is we humans who made the rules of commerce, and we can unmake them too. But we face the Sisyphean task of rethinking not just the entrenched pursuit of self-interest, the lack of reciprocity, the entitlement accorded property, but also the story of isolation that makes it all possible. When we start with the idea of isolated, selfish individuals, the job of building community becomes nearly impossible, for we understand in a place deeper than our minds that we've already assigned a different job to community, and that job has more to do with restraining individual greed than it does with loving and nurturing each member of the land-community.

And when we begin with the story of an inert and dead Earth, we cut ourselves off from the possibility of intimacy with the home that gave us birth, the only habitable planet within reach, the only home we will ever know.

◡

MY MIND DRIFTS back more than fifteen years to a camping trip that took place near the beginning of my story. A friend and I were visiting the Four

Corners region to enjoy the desert bloom of late April. The land revealed a few of its many faces—warm sunshine at midmorning, sudden afternoon thunderstorms, soft falls of nighttime snow. We savored the startling fuchsia of cactus blossoms, the feathery yellow of biscuit root. We woke to the sound of canyon wrens, their haunting tones cascading downward in the early morning light.

But the biggest revelation was the rock. It was my first trip to red rock country, and I was mute with astonishment: that so much bright beauty could be found in earth! In New Mexico we walked among the variegated hills so loved by Georgia O'Keeffe. In Chaco Canyon we followed maroon cliffs bright with paint from ancestral people a thousand years ago. But my favorite place was Canyonlands, where we climbed out of our tent every morning to a ring of lavender hills and hiked every day across wide expanses of red slickrock, guided by small cairns of stones.

I fell in love with the windswept red rock, amazed that one born to wet woodlands could feel such affinity for a parched land of rock and sky. But the time of that trip was also a time of deep mourning, and I was searching for what endures.

I found it in rock, seasoned by water and wind, existing more than a billion years in its present form. I didn't know why, but I was comforted by that land.

Stillness crept into my awareness, a stillness born of touching bedrock.

We drove home across southern Utah, winding past the submerged cliffs of Lake Powell. Just west of the lake I gazed again at the maroon-purple layers flowing past the window, dumbfounded one last time by their deep-set hues. I willed myself to remember, once I sat at home before a computer and a dissertation, how alive these cliffs were, how astoundingly red.

I found myself slipping into a deeper state of quiet, my awareness absorbed into the flowing curves of claret stone. As I watched, I began to hear, from deep inside, faint strains of music. It was like the four-part harmonies of my Mennonite childhood, only deeper, wilder—more resonant and at the same time more haunting.

"The rock is singing!" I exclaimed.

My friend looked at me, bemused.

For weeks after arriving home, I dreamed night and day of those rocks. When I sat before my computer, lavender-bright hills intruded into my sight. At night, rufous cairns beckoned me across vermilion bluffs.

~

THE THIRTEENTH-CENTURY POET Rumi wrote,

> *The speech of water, the speech of earth, and the speech of mud*
> *Are heard by those who listen with the heart. (Masnavi 1.3292)*

Rumi is often taken to mean that only mystics can hear the Earth speak—and that mystics are a strange kind of bird. But to read him that way goes against what Rumi yearned for above all—for *every* heart to be struck open by divine longing, for love to pierce every breast.

What is needed, Rumi said, is to polish the heart like a mirror.

> *Do you know why the mirror does not reflect?*
> *Because the rust is not removed from its surface. (Masnavi 1.34)*

Sufis often call this surface tarnish the "rust of otherness." Clean your mirror of all that is not love, Rumi was saying. Remember the radiance that suffuses each heart, and polish your own mirror until you can reflect it clearly. Hearing the speech of Earth may be easy when one is overcome with awe in the presence of billion-year-old radiant red rock, but it is much harder in the hubbub of the mundane.

My friend Annette recently heard the poet Gary Snyder speak. At the end of his reading, she says, a member of the audience asked Snyder how people can be inspired to save the planet. Snyder thought for a moment and said, "The planet doesn't need us to save it. The planet needs us to save ourselves. If we learned how to be better people, we would be doing good work." The roomful of activists sat in stunned silence, trying to absorb his words. Snyder went on to say, "The planet, if we notice, takes care of itself. Watch a place for a while. Look to the seasons, the weather, the animals, our own inner rhythms. Walk trails and notice things. We don't have to do a thing."

Becoming better people. It will involve remembering how to listen—to the land as well as to one another. Relearning the rhythms of give-and-take, in our own bodies as well as in our relations with others. Remembering the radically communitarian nature of life on Earth, which means remembering how to share. For however great is the divide between the very rich and the rest in this country, the gap between the industrialized nations and the rest of the world is far, far greater. The statistic is well known: less than 20 percent of the world's people are now consuming more than 80 percent of the world's resources. Anishinaabe leader Winona LaDuke says we cannot continue to use more than our share and expect to be sustainable. "You can't do that and live in accordance with natural law. That is simple logic. Most of our teachings say that."

We don't need to save the planet, but we are in desperate need of saving ourselves.

Will we learn to build an Earth-friendly culture before it is too late? Plenty of other peoples have done so, and their varied experiences offer some guidelines about what works.

Cultures that are sustainable help one another remember interconnection. They value reciprocity and fairness, and they build interdependence into their systems of exchange. They teach their children to respect others, both human and other than human. They minimize inequality among themselves, for the alternative is costly in terms of damaged health and human relationships. They observe nature closely, seeking to pattern their relationships on those of the more-than-human world. They listen to the voices of animals and plants, clouds, fish, soil, and wind, for these are relatives whose choices, along with those of humans, are in every moment creating the world. They remind themselves continually that the only way to survive and live well is to fit into the processes of the place called home—to dwell in symbiotic relationship with the land, using the gifts of Earth sparingly and taking only what is needed to live. They honor individuality among humans as part of the ongoing creative work of nature. They treasure the individuality of their place and work to preserve its unique personality, eating native foods and building their homes with nearby materials. They use local resources, yes, but first of all they love those resources as relatives. They consider themselves guests on the planet rather than owners, and so they value a mind-set of gratitude and wonder. They accept death as well as life. They shower children with love and support. They practice caring for one another and the wider land-community because love is the surest route to flourishing—and the more enjoyable way to live. They reward giving as well as taking.

For what is gathered in must be given out. What is at one time collected, another time dispersed. Breathed in, breathed out. This is the law of the ground, the law of the living Earth.

.ᔐ

THE MAIDU ELDER I was listening to finished his talk by explaining, "We begin with the Earth, then we work our way upward from there.

"It all starts with Earth."

EPILOGUE

Night

*If we had a keen vision and feeling of all ordinary human life,
it would be like hearing the grass grow and the squirrel's heart
beat, and we should die of that roar which lies on the other side
of silence.*

— GEORGE ELIOT, *MIDDLEMARCH*

I T IS 11:00 PM when I slip behind the wheel for my shift, heading east
out of Reno. The past week has been a blur of packing and scrubbing,
selling furniture on eBay, cleaning out everything we no longer need, de-
livering boxes and then more boxes to thrift store and recycling center and
dump, and scrubbing some more.

Two months ago our Grass Valley cottage finally sold, on a balmy
Sunday afternoon in mid-November when the garden glowed crimson and
gold under a sapphire sky, the cottage windows were thrown open to a
breeze that tasted of summer, and the scent of late-blooming jasmine lingered in the air. The buyers arrived, took one turn through the house, and
made an offer. Within hours the cottage was theirs.

The day after the sale, blustery cold rains clouded the skies and the jasmine settled into winter sleep.

But on this chill night in January I am thinking about more than just moving across the country and starting a new life. This afternoon, as we were saying good-bye to dear friends in Grass Valley, I received a call from Ohio. It was my brother's wife letting me know Bro's cancer had recurred. He was in the hospital again. As I listened to the message my belly lurched, the kind of intestinal heave that lets you know trouble is on the way. I didn't realize it, but I went into shock.

～

Now it is midnight. I set a tall cup of black coffee into the cup holder of the dash. It will take this much caffeine to keep me awake through the night—me who falls asleep, sitting up or not, at 10:00 PM.

But tonight I feel strangely awake. It must be the adrenaline of the last few days, the strain of saying good-bye to our California friends, the shock of hearing about my brother. Tim and I are already half a day late on our schedule to meet the moving truck in Boulder. We have to get across Nevada tonight; at the other end of it a motel that takes pets is holding a room for us. The car is packed to the gills; Brio, silent in his carrier, sits just behind the front seat. I hope he's asleep; he will weather this cross-country trip better if he can sleep. In the passenger seat Tim has already dozed off, a pillow squeezed between his head and the window.

But I feel oddly quiet and alert. In fact, I am in such a state of high alert that sleeping would be impossible even if I were lying down. Strangely, the coffee doesn't touch me either—no pounding heart, no rattled nerves. I simply feel calm. Too calm.

～

AT FERNLEY I head south off the interstate toward Highway 50. I was looking forward to showing Tim the ethereal sweeps of the Nevada desert—ridge after low ridge of bluish land resting under even bluer sky. But getting a late start means that we will cover it now in the dead of night.

The night is crystalline. Our headlights peer far down the two-lane highway. It is deserted except for the stars. Oh, the stars! In the desert-dark they are raucous with their light. They bloom overhead, their light flowing in rivers across a moonless sky. I wish I were in a convertible so I could be nearer to all those stars!

Headlights approach us only rarely. I play a game—how many miles between cars? Twenty-two miles this time. Then two cars at mile thirty-six. And then no one at all for more than fifty. Every so often my high beams light up a pair of eyes near the road, luminous and watchful.

I am grateful for the eyes, and for the stars.

~

THE CAR IS drafty. We have laid a blanket over Brio's carrier to keep him warm. Brio will keep sneezing blood in each motel room on the trip. Every morning we will wipe down sprays of blood in the bathroom before climbing into the car. For three days he will remain utterly silent in his carrier, yet when we arrive in Boulder on a snowy afternoon and carry him inside, he will spring out of his carrier and dash through our new rented house like a kitten, zipping through the rooms on the main floor, down the stairs and around the basement, then back up the stairs in one long sweep. Then he will settle on his special pillow as if he has resided in that house forever.

When we too are settled, toward the end of February, I will make a trip to Ohio to visit my brother in the hospital. He will be running a fever after a week of intensive chemo. His room will be filled with flowers, with Bro looking hearty as usual. But his skin will be pale and his eyes a little perplexed, like a child who can't understand the pain. He still brandishes an off-color joke now and then, but in between he falls into sleep, and I sit on the edge of his bed, massaging his feet or back and listening to podcasts. Blinds across the huge window are closed to slits, the room in perpetual dusk. Every so often his dozing is punctuated by abdominal pains ripping through him, and he wakes for a few moments and writhes silently, gritting his teeth, then falls back asleep. Now and then he lifts his head, surprise on his face, then finds me holding his feet. "Oh, that feels so good," he murmurs, smiling gently before dropping off again into sleep. An atmosphere of kindness enters the room with every nurse, and my brother makes a special effort to wake up and heartily thank each one.

I have not yet met the six-month-old puppy we will adopt from the humane society a week after I return home from seeing Bro. White with black spots and ears of velvety chestnut, the puppy will keep the elder cat Brio company in his final year of life. Bodhi's eyes are brown, deep and sincere, and soon he will respond with a jump to each glance from a human eye or twitch of a human muscle, attentive as the finest English butler. He will grow up to be a natural athlete, and together he and I will explore mountain trails and dog agility courses. He will also often feel afraid, especially to be left alone, which is complicated by the fact that when he has been with us only a few days, before we have even settled on a name for him, I get an emergency call from my nephew.

"Dad died this morning," my nephew will say. My brother slips away at about five on a Sunday morning, as I am dozing through a dream in which I am escorting a young teenager to a banquet. I have been paired, as a mentor, with a shy and withdrawn boy of twelve or thirteen. We sit across from each other at the long banquet table, and I try to draw him into conversation. After dinner, when it is time for me to leave, I seek him out to say good-bye, and for a moment surprised pleasure lightens his face. Only much later do I wonder about the connection between the dream and my brother.

After we get my nephew's call we look for emergency boarding for the puppy, and then Tim and I attend my brother's funeral, where the congregation, swelling with love, sings Bro's new song welcoming everyone into God's house, and one after another people stand up to share about how much my brother meant to them. The pastor speaks openly about my brother's struggles but also of his sincerity, his love for something larger than himself, for God. Marching up and down the platform of the church, the pastor will eulogize my brother, his voice rising to a triumphant pitch: "He was a flawed man"—pause—"with an aroma of holiness!" I will be shocked at the outpouring of affection from the congregation and grateful that my brother was so well loved. I will wonder if I have done half so well in finding community.

A week later I will turn fifty.

In Boulder I will feel desolate for a time, as if I have lost all the years of my childhood, for no one remains to help carry those memories. I will seek out billion-year-old rocks for comfort.

Over time the pain will ease, and I will find new joys, new friends.

ↄ

BUT ALL THAT lies in the future. For now I drive through the crisp desert night. The black earth is visible only by contrast: it is the place where the bejeweled black sky ends.

Space. I expand outward into the night, grateful for the deep stillness. Tim is asleep, Brio is silent. Just me and the stars and the night critters. Finally there is enough room. Here we are, squeezed into a packed car, but I am melting across the landscape as though I've just been let out of prison. I remember Shvana's pictures of sinking peacefully into the All. I remember

Sapphire's happy dance moments after her death. My breath slows and deepens, once again filling my body. Healing tears begin to flow.

I drive and drive, weeping, grateful, nursed by the night.

Beside the road, pairs of eyes keep company.

ACKNOWLEDGMENTS

I WILL NEVER FORGET the Korean hot-pot lunch in San Francisco in 2003 with Tina Fields and Birrell Walsh that lasted well toward dinnertime, where Birrell, hearing one of my stories about Sapphire, shoved a napkin and a pen toward my plate and said, "Write that down!" I scribbled a few words and asked, "Then what?" He said, "Keep writing!" Thank you, Birrell and Tina, for our magical conversations and for your encouragement at the start.

Neither will I forget the summer evening a short time later when, after reading the introduction to David Abram's book *The Spell of the Sensuous*, I felt my whole body buzzing as if electrified because I had just glimpsed a vision for this book and had written out its first table of contents (almost identical to the final one except I hadn't yet lived beyond chapter 9). From David I learned how to tell stories of deeply personal connection with animals and plants, and from him I caught a glimpse of how experiencing intimate ties with other creatures might be linked to the worldviews and practices of many indigenous peoples—and how such intimacies might hold a key to hope for the life of the world. Thank you, David, for imparting courage.

Many of the stories that frame each chapter were written in journals, then collected and reflected upon during weekly Keeping Writing sessions in Donna Hanelin's studio in Nevada City, California, from 2004 to 2006.

Thank you, Donna, for providing doors into writing, for your kind enthusiasm, and for patiently repeating, "Follow the image."

In January 2008 I consulted editor Dorothy Wall on my nearly complete memoir and gulped hard when the first words out of her mouth were "The most interesting part of this book hasn't been written yet." Thank you, Dorothy, for your kind and steady presence at the start while I groped my way toward a synthesis of thinking and feeling on the page, and for letting me know when the results did—and didn't—hit the right note. And for the best editorial line ever.

When I read Graham Harvey's book *Animism: Respecting the Living World* in 2006, two as-yet-unrelated parts of my life suddenly fused: a spirituality of connecting with nature and a work life in academia. Thank you, Graham, for providing an intellectual framework for the vision that became the rest of this book.

Terry Tempest Williams's *Refuge* showed me writing where I truly felt at home. Thank you, Terry, for sowing seeds that are blossoming only now, twenty years later.

I am grateful that moving to Boulder led me to the best agent ever, Kristina Holmes. Thank you, Kristina, for believing in the book from the moment you picked it up, for your consummate professionalism, and for communicating with me so regularly and well while representing the book. And then for becoming a friend.

To the Boulder Media Women, thank you for bringing Kristina and me together at a Tuesday afternoon schmooze and for all the scrumptious potlucks, intriguing conversations, and good-hearted support flowing throughout the group.

My colleagues and students at Prescott College teach me something in every conversation. Thank you, all of you, for loving our dear Earth and being committed to helping others remember how to love those who dwell here.

Each of the authors of the several hundred books I edited over more than twenty-five years showed me something about writing. Thank you for providing ten thousand hours of writing practice, and especially for teaching me to enjoy polishing prose.

During the early years of writing this book, I benefited from writing workshops at Prescott College led by Susan Griffin, Demetria Martinez, Rick Bass, and Craig Childs. In the middle of the book I enjoyed a New Year's weekend writing retreat with Marilyn Krysl and Sandra Dorr. Thank you, Marilyn and Sandy, for your evocative writing prompts and your heart-opening late-night readings, and to you, Sandy, for advising me to write a sentence that fills a page. I haven't yet managed it, but I hope to someday.

A number of friends and colleagues helped me with research questions or read sections of the manuscript and offered comments. Thank you to the editor Mimi Kusch for being my earliest reader and to the editor Jody Berman for being one of the most recent. Thanks as well to Barbara Voss, Christopher Brown, Gail Storey, Gary Holthaus, James Pittman, Jane Compson, Jared Aldern, Judith Trent, Julene Bair, Linda Hogan, Lori Stott, Marilyn Krysl, Marj Hahne, M. J. Barrett, M. J. Zimmerman, Robin Hammer, Sandra Dorr, and Timothy Falb. I am grateful to Carol A. Wilson and the Stillpoint Center for hosting the first public reading from the book. Thank you to a writing group made up of five wonderful writers, Elisabeth Hyde, Gail Storey, Julene Bair, Lisa Jones, and Marilyn Krysl, who all together said, "We have to see what difference this idea makes to *you*." I took your words to heart.

When all was said and nearly done, Mimi Kusch read the entire manuscript and suggested the tiniest, most graceful wording changes ever. Over and over, Mimi, you were right. Thank you for knowing me, and now the book, so well.

Two librarians at Prescott College scrambled to find obscure interlibrary loan articles as I plowed my way through environmental history, economics and religion, and Mesopotamian social structures. Bill Fiscus, you found the vast majority, and Helen Manion, be glad you didn't come on board until the book was nearly finished! Thank you both for your persistence and prompt service.

The bulk of the book was written in a Boulder house where my computer sat beside a window looking out on open space, and in between paragraphs I grabbed binoculars to watch birds. Pauline Kenny and Stephen Cohen, thank you for making available a house with skylights where I woke every morning, looked around, and thought, "I'm so lucky!"

Working with Counterpoint has been a dream come true. Jack Shoemaker, I will be forever grateful that you took a chance on an unknown writer. Julie Pinkerton, thank you for your thoughtful editorial review of the manuscript and for so gently letting me know where stories or ideas needed clarifying. I am grateful to Kirsten Janene-Nelson for finding and inserting all those subtitles and especially for saving me from some inexcusable misspellings. Ann Weinstock designed the best book cover ever using a photograph from Don Johnston. The beautiful interior design is the work of Megan Jones. Thank you to the Counterpoint production and promotion staff, including Kelly Winton, Emma Cofod, Liz Parker, Jodi Hammerwold, and Maren Fox, for expertly shepherding both the book and me out into the world.

To the friends who sweetened my life with their tree-love or dog-love or flower-love or animal-love or rock-love: thank you for conveying hope. You all know who you are. I owe my life to you. We all do.

Sapphire, you were my partner for ten of the years covered in this book. You alone know how much you contributed to the book's unfolding, and to mine.

Finally, to my sweetheart, Tim: thank you for believing wholeheartedly in my vision and for giving me all these years of steady, sure-footed love. And for filling in the nouns when I needed them.

NOTES

PROLOGUE: BALD EAGLE

The chapter epigraph comes from Annie Dillard, *Teaching a Stone to Talk: Expeditions and Encounters* (New York: HarperCollins, 1982), 88.

Carol Lee Flinders's lovely phrase "fissures in ordinary logic" appears in *At the Root of This Longing: Reconciling a Spiritual Hunger and a Feminist Thirst* (San Francisco: HarperSanFrancisco, 1998), 14, a book I had the pleasure of copyediting.

Thomas S. Kuhn's influential book is *The Structure of Scientific Revolutions*, 2nd ed. (Princeton: Princeton University Press, 1970); quote is found on 94.

For empathy in mice, see Dale J. Langford, Sara E. Crager, Zarrar Shehzad, Shad B. Smith, Susana G. Sotocinal, Jeremy S. Levenstadt, Mona Lisa Chanda, Daniel J. Levitin, and Jeffrey S. Mogil, "Social Modulation of Pain as Evidence for Empathy in Mice," *Science* 312, no. 5782 (2006): 1967–70.

Ursula K. Le Guin's lovely quote about the dreaming of rocks comes from *The Lathe of Heaven* (1971; New York: Scribner, 1999), 167.

Michael Pollan outlines the stories of the tulip, potato, apple, and marijuana in *The Botany of Desire: A Plant's-Eye View of the World* (New York: Random House, 2002).

Anthropologists of religion have been leading the way in redefining animism from religion that focuses on "spirits" to religion that focuses on "persons," who may have either material or more subtle bodies. The Northern Ojibwe man's comment about stones is found in A. Irving Hallowell, "Ojibwa Ontology, Behaviour, and World View," in *Culture in History*, ed. Stanley Diamond (New York: Columbia University Press, 1960). Nurit Bird-David, an Israeli anthropologist, uses Hallowell to redefine *animists* as people who use relationships to understand their environment; see "'Animism' Revisited: Personhood, Environment, and Relational Epistemology," *Current Anthropology* 40 (1999): 67–79. Both of these essays and others on similar themes are reprinted in Graham Harvey, *Readings in Indigenous*

Religions (New York: Continuum, 2002). Harvey drew these ideas together in his book *Animism: Respecting the Living World* (New York: Columbia University Press, 2005). The phrase about relating to people, "only some of whom are human," comes from Harvey. David Abram's influential and popular book is *The Spell of the Sensuous: Perception and Language in a More-Than-Human World* (New York: Pantheon, 1996); quote is from 131. Abram's more recent book is *Becoming Animal: An Earthly Cosmology* (New York: Random House, 2010).

The scholar who has done most to articulate and question American notions of the wilderness is William Cronon, best known for his book *Changes in the Land: Indians, Colonists, and the Ecology of New England* (New York: Hill and Wang, 1983). His 1995 *New York Times* essay, "The Trouble with Wilderness," is available at www.williamcronon.net/writing/Trouble_with_Wilderness_Main.html.

The term *avianomorphism* comes from Judith Irving, *The Wild Parrots of Telegraph Hill*, Independent Lens, www.pbs.org/independentlens/wildparrots/qa.html.

1: CUT-LEAF WEEPING BIRCH

Snippets of Dorothy Maclean's conversations with plants are found in the book I edited, *To Honor the Earth: Reflections on Living in Harmony with Nature*, with photographs by Kathleen Thormod Carr (San Francisco: HarperSanFrancisco, 1991). For more on Maclean's experiences at Findhorn Gardens in northern Scotland, see her memoir, *To Hear the Angels Sing: An Odyssey of Co-Creation with the Devic Kingdom* (Hudson, NY: Lindisfarne Press, 1990).

The words of Audre Lorde on flesh and reason are from her poem "On a Night of the Full Moon," in *Chosen Poems Old and New* (New York: Norton, 1982). Her experiences with cancer and racism are found in her journals, published as Audre Lorde, *A Burst of Light: Essays* (Ithaca, NY: Firebrand Books, 1988).

Eknath Easwaran's translation of the Upanishads is warm and accessible and includes some helpful introductions to the texts by Michael N. Nagler; see Eknath Easwaran, trans., *The Upanishads*, 2nd ed. (Tomales, CA: Nilgiri Press, 2007). The lines beginning "As the same air" are from Katha Upanishad 2.2.9, and "Eternal peace is theirs" is from Katha Upanishad 2.2.13, both on p. 88.

The writings of anthropologist Eduardo Viveiros de Castro are tough going but well worth the effort; the ideas I discuss here are found in "Cosmological Deixis and Amerindian Perspectivism," *Journal of the Royal Anthropological Institute*, n.s., 4, no. 3 (1998): 469–88.

Soul as "a disembodied entity hidden within the outer shell of the disposable body" comes from N. T. Wright's book *Surprised by Hope: Rethinking Heaven, the Resurrection, and the Mission of the Church* (San Francisco: HarperOne, 2008), 28. Wright argues that modern-day Christians are more Platonic than Christians of the early church; it was a simple step from there to suggesting that modern people are more Platonic (in the sense of seeing a radical difference between soul and body) than Plato himself.

The book *Psyche and Soma: Physicians and Metaphysicians on the Mind-Body Problem from Antiquity to Enlightenment*, ed. John P. Wright and Paul Potter (Oxford: Clarendon Press, 2000), contains a wealth of information on views of mind and body throughout Western history. For information on Plato, I relied on T. M. Robinson's chapter, "The Defining Features of Mind-Body Dualism in the Writings of Plato." For later periods I used Emily Michael's chapter, "Renaissance Theories of Body, Soul, and Mind," and Stephen Voss's "Descartes: Heart and Soul."

The outline of the great chain of being comes from Arthur O. Lovejoy, *The Great Chain of Being: A Study of the History of an Idea* (1936; reprint, Cambridge, MA: Harvard University Press, 1960).

Thomas Berry may be best known for his collaboration with cosmologist Brian Swimme in *The Universe Story: From the Primordial Flaring Forth to the Ecozoic Era—A Celebration of the Unfolding of the Cosmos* (San Francisco: HarperSanFrancisco, 1992). Anything by Berry will open one's heart toward friendlier relations with the Earth, and I have benefited greatly from *The Great Work: Our Way into the Future* (New York: Bell Tower, 1999), and *Evening Thoughts: Reflecting on Earth as Sacred Community* (San Francisco: Sierra Club Books, 2006). The Berry quote comes from *The Sacred Universe: Earth, Spirituality, and Religion in the Twenty-first Century* (New York: Columbia University Press, 2009), 173.

On mind, body, and emergentism, see John Heil, *Philosophy of Mind: A Contemporary Introduction*, 2nd ed. (New York: Routledge, 2004), and Brian P. McLaughlin and Jonathan Cohen, eds., *Contemporary Debates in Philosophy of Mind* (Malden, MA: Blackwell, 2007). The alternative to emergentism is panpsychism, or the idea, which I am arguing for here, that matter is imbued with mind; for a helpful introduction to panpsychism, see William Seager and Sean Allen-Hermanson, "Panpsychism" (2005), in the online Stanford Encyclopedia of Philosophy, http://plato.stanford.edu/entries/panpsychism.

The meditation exercise to connect with a tree that I engaged in with the cottonwood on the banks of Granite Creek in Prescott, Arizona, was inspired by a meditation for connecting with animals taught by my friend Winterhawk, by a yogic meditation for opening the lens of perception, and by the cottonwood tree.

Einstein's words on a sense of awe leading to discovery are quoted in Evelyn
Fox Keller's biography of Barbara McClintock, *A Feeling for the Organism: The
Life and Work of Barbara McClintock* (New York: W. H. Freeman, 1983). Much of
my information on McClintock comes from this book. Stephen Jay Gould's take on
McClintock is found in his review of Keller's biography, "Triumph of a Naturalist,"
New York Review of Books 31, no. 5 (March 1984). Another, more critical, book on
McClintock is by Nathaniel C. Comfort, *The Tangled Field: Barbara McClintock's
Search for the Patterns of Genetic Control* (Cambridge, MA: Harvard University
Press, 2001). Comfort was helpful in clarifying that McClintock's "most mystical-
sounding ideas stemmed from observation and skepticism, not occult visitations"
(267), but he views her as not having had an intimate partnership in her life. The
Science writer referred to is Gene E. Robinson, "Beyond Nature and Nurture,"
Science 304 (April 16, 2004): 397–99.

Malidoma Patrice Somé's early book, *Of Water and the Spirit: Ritual, Magic
and Initiation in the Life of an African Shaman* (New York: Penguin, 1995), details
his initiation rituals into nature and is guaranteed to stretch the limits of the reader's
thinking about nature. The Dagara philosophical system is outlined in a more systematic way in *The Healing Wisdom of Africa* (New York: Tarcher, 1999), for
which I served as development editor.

2: HOODED ORIOLE

The Carol Lee Sanchez quote comes from "Animal, Vegetable, Mineral: The Sacred
Connection," in *Ecofeminism and the Sacred*, ed. Carol J. Adams (New York:
Continuum, 1993), 225.

Jung's technique of active imagination received a lot of attention among ther-
apists and clients in the late eighties and early nineties as a result of Robert A.
Johnson's still-popular book *Inner Work: Using Dreams and Active Imagination for
Personal Growth* (New York: Harper & Row, 1986). For a collection of Jung's writ-
ings on the subject, see *Jung on Active Imagination*, ed. Joan Chodorow (Princeton:
Princeton University Press, 1997).

A concise summary of Jung's understanding of archetypes is found in Michael
Vannoy Adams, "The Archetypal School," in *The Cambridge Companion to
Jung*, ed. Polly Young-Eisendrath and Terence Dawson (Cambridge: Cambridge
University Press, 2008), 107–8. That Jung argued for a physical basis for the ar-
chetypes is clear: "People still think that relationships like this are far-fetched and
therefore improbable. But they forget that the structure and function of the bodily

organs are more or less everywhere the same, including those of the brain. And as the psyche is to a large extent dependent on this organ, presumably it will—at least in principle—everywhere produce the same forms"; see C. G. Jung, *The Collected Works of C. G. Jung*, trans. R. F. C. Hull (New York: Pantheon, 1963), 14:xix.

Demaris S. Wehr's respectfully critical analysis of Jung appears in *Jung and Feminism: Liberating Archetypes* (Boston: Beacon Press, 1989). Barbara Alice Mann's words are found in *Iroquoian Women: The Gantowisas*, American Indian Studies 4 (New York: Peter Lang, 2000), 62.

Jung's story of the patient with the dream of the golden scarab beetle is found in C. G. Jung, "On Synchronicity," *Collected Works*, trans. R. F. C. Hull, 2nd ed. (Princeton: Princeton University Press, 1969), 8:525–26. Jung's dialogues with the inner figure who recommended that thoughts be treated "like animals in the forest" appears in *Memories, Dreams, Reflections*, ed. Aniela Jaffé (New York: Random House, 1963), 183; his vision of war appears on 176. David Abram's quote is from *The Spell of the Sensuous* (New York: Vintage, 1996), 22. Jung's words about isolation come from *Memories, Dreams, Reflections*, 194–95.

Vine Deloria Jr.'s thoughts on relationship, place, and a personal universe come from "Power and Place Equal Personality," chap. 3 in *Power and Place: Indian Education in America*, by Vine Deloria Jr. and Daniel R. Wildcat (Golden, CO: American Indian Graduate Center and Fulcrum Press, 2001), 23.

3: WILD ORPHANS

Wendell Berry's poems "To Know the Dark" and "Song in a Year of Catastrophe" appear in *Farming: A Hand Book* (Berkeley, CA: Counterpoint, 2011).

A good source for statistics on deforestation in Thailand as well as a portal to information on the ecological monks of Thailand is EcoLocalizer, http://ecolocalizer .com/2008/01/19/thai-monks-combat-deforestation. For an in-depth discussion of the monks, see Susan M. Darlington, "Rethinking Buddhism and Development: The Emergence of Environmentalist Monks in Thailand," *Journal of Buddhist Ethics* 7 (2000). The Religion and Ecology portal at Yale includes a story on the monks, with more sources: http://fore.research.yale.edu/religion/buddhism/projects/thai_ecology .html.

Christians for the Mountains can be found at http://christiansforthe mountains.org. Their website includes resolutions against mountaintop removal from all the major Christian denominations. An article on Allen Johnson from *Grist* is at www.grist.org/article/johnson2. For a scholarly study of faith-based projects in Appalachia, see David Lewis Feldman and Lyndsay Moseley, "Faith-Based Environmental Initiatives in Appalachia: Connecting Faith, Environmental Concern and Reform," *Worldviews* 7, no. 3 (2003): 227–52.

4: RED FOXES

The book on tracking in which I found information about red foxes is by Donald and Lillian Stokes, *Stokes Guide to Animal Tracking and Behavior* (Boston: Little, Brown, 1986), 286.

On the history of Mennonites and Amish, the Global Anabaptist Mennonite Encyclopedia Online, www.gameo.org, has a wealth of information. In writing this brief section I consulted the entries "Alsace (France)," "Ammann, Jakob (17th/18th Century)," "Amish Division," "Amish Mennonites," "Bern (Switzerland)," "Ban," and "Marital Avoidance."

The quote from Tobias Wolff comes from his essay "Reconsidering Paul Bowles," Narrative Magazine, www.narrativemagazine.com. In a synchronicity, his essay crossed my desk for copyediting just as I was beginning to write this chapter.

The story of predator killing in the twentieth century is told in many books on wildlife history. Statistics on coyote killing come from Thomas R. Dunlap, *Saving America's Wildlife: Ecology and the American Mind* (Princeton: Princeton University Press, 1988), chap. 4. At the beginning of the all-out campaign, in 1915, Congress appropriated $125,000 to train hunters in killing wolves. By the early 1940s the budget for eliminating predators was $3 million, and by 1971 the combined federal-state budget for it was $8 million; see Donald Worster, *Nature's Economy: A History of Ecological Ideas*, 2nd ed. (Cambridge: Cambridge University Press, 1994), 263–64. For a summary of public reaction to the policy, see James W. Feldman, "Public Opinion, the Leopold Report, and the Reform of Federal Predator Control Policy," *Human-Wildlife Conflicts* 1, no. 1 (Spring 2007): 112–24. Up-to-date statistics on predator elimination policies are available from the Oregon-based nonprofit Predator Defense. Their page "The USDA's War on Wildlife" includes history, charts, and details on poisons used: www.predatordefense.org/USDA.htm.

The short quote by Thomas King is from *The Truth About Stories: A Native Narrative* (Toronto: House of Anansi Press, 2003), 146.

The writing on Aldo Leopold is voluminous and fascinating. Two excellent biographies are: Susan L. Flader, *Thinking Like a Mountain: Aldo Leopold and the Evolution of an Ecological Attitude toward Deer, Wolves, and Forests* (Madison: University of Wisconsin Press, 1974), and Curt Meine, *Aldo Leopold: His Life and*

Work (Madison: University of Wisconsin Press, 1988). Details of the wolf's death come from Meine's biography. The influence of P. D. Ouspensky on Leopold is discussed by Flader on p. 18 and Meine on pp. 214–15; the quote from Ouspensky is found in Meine, 215. The quote beginning "Possibly, in our intuitive perceptions" is from Leopold's "Some Fundamentals of Conservation in the Southwest" (1923), quoted in Meine, 214, and available on the website of the Aldo Leopold Foundation, www.aldoleopold.org/about/outlook/winter2008/fundamentals.shtml. Quotes from Aldo Leopold's *Game Management* (1933) come from Worster, *Nature's Economy*, 272. "Thinking Like a Mountain" and "The Land Ethic" come from *A Sand County Almanac* (1949; reprint, London: Oxford University Press, 1968).

J. Baird Callicott has written a great deal on Leopold's land ethic, including *In Defense of the Land Ethic: Essays in Environmental Philosophy* (Albany: State University of New York Press, 1989) and "Aldo Leopold," in *Fifty Key Thinkers on the Environment*, ed. Joy A. Palmer (London: Routledge, 2001), 174–80. Marti Kheel critiques Leopold and environmental ethicists in general for emphasizing the biotic community at the expense of the welfare of individual animals in *Nature Ethics: An Ecofeminist Perspective* (Lanham, MD: Rowman & Littlefield, 2008).

On the topic of biodiversity, I find these books helpful: Vandana Shiva, Patrick Anderson, Heffa Schücking, Andrew Gray, Larry Lohman, and David Cooper, *Biodiversity: Social and Ecological Perspectives* (London: Zed Books, 1991); Joel B. Hagen, *An Entangled Bank: The Origins of Ecosystem Ecology* (New Brunswick, NJ: Rutgers University Press, 1992); Charles C. Mann and Mark L. Plummer, *Noah's Choice: The Future of Endangered Species* (New York: Knopf, 1995); David Takacs, *The Idea of Biodiversity: Philosophies of Paradise* (Baltimore: Johns Hopkins University Press, 1996); Timothy J. Farnham, *Saving Nature's Legacy: Origins of the Idea of Biological Diversity* (New Haven, CT: Yale University Press, 2007).

The California 1991 memorandum of understanding that set the state policy as managing for ecosystem health is online at: http://biodiversity.ca.gov/mou.html.

George J. Sefa Dei's ideas about education can be found in "Afrocentricity: A Cornerstone of Pedagogy," *Anthropology and Education Quarterly* 25, no. 1 (March 1994), quote from p. 20.

Augustine's journal, titled *Confessions*, feels surprisingly contemporary because of its emotional immediacy and deserves to be on the required reading list of every survey course in Western literature. To read it in the context of the issues and movements swirling at the time, check out the deeply engaging biography of Augustine by the great scholar of late antiquity, Peter Brown: *Augustine of Hippo: A Biography*, rev. ed. (Berkeley and Los Angeles: University of California Press,

2000). Information and original writings of Pelagius can be found in: B. R. Rees, ed., *The Letters of Pelagius and His Followers* (Woodbridge, Suffolk, UK: Boydell, 1991); and B. R. Rees, *Pelagius: A Reluctant Heretic* (Woodbridge, Suffolk, UK: Boydell, 1988).

The Yurok man's teaching is related by Thomas Buckley, "Doing Your Thinking," in *I Become Part of It: Sacred Dimensions in Native American Life*, ed. D. M. Dooling and Paul Jordan-Smith (San Francisco: HarperSanFrancisco, 1992), 43.

After I had finished writing this chapter I discovered the work of anthropologist Marshall Sahlins, who decades ago had accomplished what I was now trying to do: call attention to how central the idea of original sin became in European and then American notions of politics, law, and nature. Sahlins synthesized these themes in "The Sadness of Sweetness: The Native Anthropology of Western Cosmology," *Current Anthropology* 37, no. 3 (June 1996): 395–415. His earlier book, *The Use and Abuse of Biology: An Anthropological Critique of Sociobiology* (Ann Arbor: University of Michigan Press, 1976), is also useful, especially the final chapter. My own scholarly paper on these issues is "The Animal versus the Social: Rethinking Individual and Community in Western Cosmology," in the *Handbook of Contemporary Animism*, ed. Graham Harvey (forthcoming, 2013).

5: STORIES WE LIVE BY

The epigraph from Thomas King is a recurring refrain in *The Truth About Stories: A Native Narrative* (Toronto: House of Anansi Press, 2003).

Quotes from Kathleen Dean Moore come from *The Pine Island Paradox: Making Connections in a Disconnected World* (Minneapolis: Milkweed Editions, 2004), 4.

Linda Hogan's observations on healing are found in *The Woman Who Watches Over the World* (New York: Norton, 2001), 21.

Francis Bacon's view that knowledge and power are synonyms is found in *Novum Organum* (1620; London: Routledge, 1893), bk. 1, aphorisms 3, 129.

Quotes from Thomas Hobbes are from *Leviathan*, ed. A. R. Waller (Cambridge: Cambridge University Press, 1903), 63, 84. Religious historians have argued for many years that Hobbes was an implicit Calvinist, though Helen Thornton argues that Hobbes's views were actually a bit closer to Lutheranism than to Calvinism; see Thornton, *State of Nature or Eden? Thomas Hobbes and His Contemporaries on the Natural Condition of Human Beings* (Rochester, NY: University of Rochester Press,

2005). On Hobbes as a Calvinist, see A. P. Martinich, *The Two Gods of Leviathan: Thomas Hobbes on Religion and Politics* (Cambridge: Cambridge University Press, 1992); and Philip S. Gorski, *The Disciplinary Revolution: Calvinism and the Rise of the State in Early Modern Europe* (Chicago: University of Chicago Press, 2003). Recent takes on Hobbes come from Garrath Williams, "Hobbes: Moral and Political Philosophy," Internet Encyclopedia of Philosophy, www.iep.utm.edu/hobmoral, and Scott Horton, "Hobbes—How We Make the Future from the Past," *Harper's Magazine*, August 1, 2009.

Duncan K. Foley has written an engaging book for nonspecialists on the ideology of classical economics: *Adam's Fallacy: A Guide to Economic Theology* (Cambridge, MA: Harvard University Press, 2006). In it he draws attention to the fallacious assumption in classical and neoclassical economics that the economic sphere of life can be separated from the social. Amartya Sen clarifies Smith's position in *On Ethics and Economics* (Oxford: Basil Blackwell, 1987), 23–25.

Biologists too are reaching toward less-reductionistic stories than they told in the past. On symbiosis as a mechanism of evolution, see Frank Ryan, *Darwin's Blind Spot: Evolution Beyond Natural Selection* (Boston: Houghton Mifflin, 2002); Luis P. Villarreal, "Can Viruses Make Us Human?" *Proceedings of the American Philosophical Society* 148, no. 3 (September 2004): 296–323; and Luis P. Villarreal, *Viruses and the Evolution of Life* (Washington, DC: American Society for Microbiology, 2005). Robert Wesson, in *Beyond Natural Selection* (Cambridge, MA: MIT Press, 1993), reviews the debate about natural selection and emphasizes the role of cooperation in evolution.

For the thought of Thomas Robert Malthus, I enjoyed John Avery, *Progress, Poverty, and Population: Rereading Condorcet, Godwin, and Malthus* (London: Frank Cass, 1997), and parts of Eric B. Ross, *The Malthus Factor: Poverty, Politics, and Population in Capitalist Development* (London: Zed Books, 1998), and Robert M. Young, *Darwin's Metaphor: Nature's Place in Victorian Culture* (Cambridge: Cambridge University Press, 1985), 73. The quote from Malthus comes from Malthus's *An Essay on the Principle of Population*, 3.2.3, www.econlib.org/library/Malthus/malPlong13.html.

Donald Worster's account of Darwin's discoveries is evocative and engaging: *Nature's Economy*, 2nd ed. (Cambridge: Cambridge University Press, 1994); Darwin's quote about Staffordshire is found on 119 and Melville's on 121. Darwin recorded his encounter with Malthus in *The Life and Letters of Charles Darwin*, vol. 1, ed. Francis Darwin (orig. publ. 1887), Gutenberg Project, www.gutenberg.org/cache/epub/2087/pg2087.html. Darwin's acceptance of Malthus has been discussed at length by historians and anthropologists of science. Most engaging is Worster,

Nature's Economy, 149. Robert M. Young fills out the philosophy behind the ideas in *Darwin's Metaphor: Nature's Place in Victorian Culture* (Cambridge: Cambridge University Press, 1985), especially 80–88. Peter J. Bowler argues that Malthus and Darwin's ideas of the struggle for existence were somewhat different; see "Malthus, Darwin, and the Concept of Struggle," *Journal of the History of Ideas* 37, no. 4 (October–December 1976): 631–50. Darwin's musing about the war of nature in his own backyard is quoted in Worster, *Nature's Economy*, 128.

The Pavlov quote is from Richard D. Ryder, *Animal Revolution: Changing Attitudes Toward Speciesism*, rev. ed. (Oxford: Berg, 2000), 160. Frans de Waal tells of the skeptical response by scientists to Barbara Smuts's idea of friendship among baboons in *Good Natured: The Origins of Right and Wrong in Humans and Other Animals* (Cambridge, MA: Harvard University Press, 1996), 19.

The Genesis story of the fall could not by itself have generated the world-denying attitude of the West, as Lynn White Jr. argued in "The Historical Roots of Our Ecological Crisis," *Science* 155 (March 1967): 1203–7. Genesis had to be read through the separation of matter and spirit that informed the Greek-influenced worldview of late antiquity, where salvation meant escape from the material world. Augustine and most Christians as well as most people in the empire accepted the idea of a vast distance between the earthly and the divine. "I asked the earth," Augustine wrote, "And what is this God?" It answered, "'I am not he'; and everything in the earth made the same confession. . . . I asked the heavens, the sun, moon, and stars; and they answered, 'Neither are we the God whom you seek'" (*Confessions*, 10.6.9, Christian Classics Ethereal Library, www.ccel.org/ccel/augustine/confessions.xiii .html). Jews, who originated the story of the fall, were not as thoroughly Hellenized as early Christians during those centuries of the Roman Empire and so did not absorb the Greek nature-spirit divide to the extent that Christians did. Jews have never been preoccupied with the curse of nature in the way that Christians, under the influence of Augustine, have been. Jewish theology tends to emphasize the good-ness of creation and of sex—the very things Augustine's influence cast a chill across throughout Christian history.

The Marshall Sahlins quote comes from *The Use and Abuse of Biology: An Anthropological Critique of Sociobiology* (Ann Arbor: University of Michigan Press, 1976), 100.

Quotes from the Mundaka Upanishad are my paraphrases of 3.1.1–4, based on several well-known translations. I especially like that of Eknath Easwaran, *The Upanishads* (Tomales, CA: Nilgiri Press, 2007), 192–93, and of Swami Gambhirananda, *Eight Upanishads* (Calcutta: Advaita Ashrama, 1982), 2:143–50.

That the Self can be found even in "a clump of grass" comes from the commentary of Sankaracarya found in Gambhirananda's translation (2:148).

My discussion of Andean culture is based on several essays appearing in *The Spirit of Regeneration: Andean Culture Confronting Western Notions of Development*, ed. Frédérique Apffel-Marglin with PRATEC (London: Zed Books, 1998): Julio Vallodolid Rivera, "Andean Peasant Agriculture: Nurturing a Diversity of Life in the *Chacra*," 51–88; Grimaldo Rengifo Vásquez, "The *Ayllu*," 89–123; Greta Jimenez Sardon, "The Aymara Couple in the Community," 146–71; Grimaldo Rengifo Vásquez, "Education in the Modern West and in Andean Culture," 172–93; Eduardo Grillo Fernandez, "Development or Decolonization in the Andes?" 194–243. The authors are Western-educated indigenous people who became sociologists and anthropologists and formed a nonprofit promoting indigenous Andean solutions to poverty and colonization. More information on PRATEC is available at the UNESCO site: http://creativecontent.unesco.org/media-library. The quote about the "*chacra* that walks" is from Rivera, "Andean Peasant Agriculture," 56. The quote that life "emerges from conversations between similar and equivalent beings" is from Rengifo Vásquez, "*Ayllu*," 97. "With them we keep company, with them we converse and reciprocate" is from Grillo Fernandez, "Development or Decolonization?" 221. "Conversation is thus an attitude" is from Rengifo Vasquez, "*Ayllu*," 107.

While the Andeans, in placing rights of nature into national law, are implementing an indigenous philosophy, they are borrowing the legal model from local communities in the United States that in recent years have passed local community bills of rights to recognize the rights of natural communities to exist and thrive and in cases of degradation to be restored. The Community Environmental Legal Defense Fund, based in Pennsylvania, consults with local communities on writing such legislation; see their website at www.celdf.org. Another legal initiative for promoting the rights of nature is spearheaded in Britain by the barrister Polly Higgins, who is working at the United Nations to recognize ecocide as the fifth crime against peace, alongside genocide and human rights abuses; see the website of Polly Higgins: www .pollyhiggins.com. Cormac Cullinan, attorney from South Africa, has written the most complete theoretical guide to the rights of nature; see *Wild Law: A Manifesto for Earth Justice* (White River Junction, VT: Chelsea Green Publishing, 2011).

Darwin's thoughts on being "netted together" as "fellow brethren" are as follows: "if we choose to let conjecture run wild then animals our fellow brethren in pain, disease death & suffering, & famine; our slaves in the most laborious works, our companions in our amusements. they may partake, from our origin, in one common ancestor we may be all netted together"; see Paul H. Barrett, Peter J. Gautrey, Sandra Herbert, David Kohn, and Sydney Smith, eds., *Charles Darwin's Notebooks,*

1836–1844, Transmutation of Species, Notebook B (1837–38), ed. David Kohn (Cambridge: Cambridge University Press, 2008), 228–29. Darwin elaborated on the similar instincts and feelings between humans and other social animals in *The Descent of Man: The Concise Edition*, selected by Carl Zimmer (New York: Penguin Group, 2007), 127–28. Darwin's adventures with earthworms are beautifully summarized by Eileen Crist, "The Inner Life of Earthworms: Darwin's Argument and Its Implications," in *The Cognitive Animal: Empirical and Theoretical Perspectives on Animal Cognition*, ed. Marc Bekoff, Colin Allen, and Gordon M. Burghardt (Cambridge, MA: MIT Press, 2002), 3–8.

Many animal scientists and psychologists today are documenting a departure from the mechanistic story of animals. They include: Frans de Waal, *Good Natured: The Origins of Right and Wrong in Humans and Other Animals* (Cambridge, MA: Harvard University Press, 1996), also Frans de Waal, "Are We in Anthropodenial?" *Discover* (July 1997), http://discovermagazine.com/1997/jul/are weinanthropod1180; Marc Bekoff and Jessica Pierce, *Wild Justice: The Moral Lives of Animals* (Chicago: University of Chicago Press, 2009); Paul Ekman, "Darwin's Compassionate View of Human Nature," *Journal of the American Medical Association* 303, no. 6 (February 10, 2010), available at www.paulekman.com/pub lications/journal-articles-book-chapters; also see Ekman's "Survival of the Kindest," in the November 2010 issue of *Shambhala Sun*; Dacher Keltner, "Darwin's Touch: Survival of the Kindest," *Psychology Today* blog, February 11, 2009, www.psy chologytoday.com/blog/born-be-good/200902/darwins-touch-survival-the-kindest.

For the circle of creation, see Paula Gunn Allen, *The Sacred Hoop: Recovering the Feminine in American Indian Traditions* (Boston: Beacon Press, 1990). For the "continuous web of creation" of the Maori, see Makere Stewart-Harawira, "Cultural Studies, Indigenous Knowledge, and Pedagogies of Hope," *Policy Futures in Education* 3, no. 2 (2005): 155. For a perceptive analysis of an animist cosmos as a "meshwork," a "tissue of trails that together make up the texture of the life-world," see the anthropologist Tim Ingold, "Rethinking the Animate, Re-animating Thought," *Ethnos* 71, no. 1 (2006): 9–20.

Quantum physics is often used willy-nilly to explain every strange phenomenon in nature, partly because its philosophical implications have not been spelled out by many physicists themselves. Two who do so are Arthur Zajonc and Karen Barad. A lovely interview with Zajonc was done in 2010 by Krista Tippett for the radio program *On Being*; the segment is "Holding Life Consciously," http://being.pub licradio.org/programs/2010/holding-life-consciously. Karen Barad's dense book is *Meeting the Universe Halfway: Quantum Physics and the Entanglement of Matter and Meaning* (Durham, NC: Duke University Press, 2007). I summarized a small piece of it in my paper "Being Known by a Birch Tree: Animist Refigurings of Western Epistemology," *Journal for the Study of Religion, Nature, and Culture* 4, no. 3 (September 2010): 182–205.

6: SAPPHIRE

For two different views of how dominance theory came to dominate dog training, see Morgan Spector, "Moving Beyond the Dominance Myth," available at 4 Paws University, www.4pawsu.com/dogpsychology.htm, and David Ryan, "Why Won't Dominance Die?" Association of Pet Behaviour Counsellors, www.apbc.org .uk/articles/why-wont-dominance-die. For an accessible summary of the issues and citations to recent animal research, see Debra Milikan, "The Alpha Theory: Based on a Misguided Premise," C.L.E.A.R Dog Training, www.cleardogtraining.com.au /articles/training-articles/105.

A concise outline of wolf culture based on a dozen years of observing wolves in the wild is found in L. David Mech, "Alpha Status, Dominance, and Division of Labor in Wolf Packs," *Canadian Journal of Zoology* 77, no. 8 (August 1999): 1196–1203. Mech explains the misuse of the term *alpha wolf* in a YouTube video: www .youtube.com/watch?v=tNtFgdwTsbU. Two studies reviewing the various definitions of dominance in animal behavior literature are: Carlos Drews, "The Concept and Definition of Dominance in Animal Behaviour," *Behaviour* 125, nos. 3–4 (1993): 283–313; and James O'Heare, "Social Dominance: Useful Construct or Quagmire?" *Journal of Applied Companion Animal Behavior* 1, no. 1 (2007): 56–83. The research and publications of animal scientists Marc Bekoff, Ray and Lorna Coppinger, and Wendy van Kerkhove all provide helpful alternatives to dominance theory.

For the experiment measuring dog, wolf, and infant responsiveness to human communications, see Jozsef Topal, György Gergely, Ágnes Erdhegyi, Gergely Csibra, and Ádám Miklósi, "Differential Sensitivity to Human Communication in Dogs, Wolves, and Human Infants," *Science* 325, no. 5945 (September 4, 2009): 1269–72.

The quote from Nicholas Dodman, a director at the Tufts Animal Behavior Clinic of the Cummings School of Veterinary Medicine at Tufts University, comes from Anna Bahney, "C'mon, Pooch, Get with the Program," *New York Times*, February 23, 2006, www.nytimes.com/2006/02/23/fashion/thursdaystyles/23dogs .html.

On humans winning status by being generous to others, see Cameron Anderson and Gavin J. Kilduff, "The Pursuit of Status in Social Groups," *Current Directions in Psychological Science* 18, no. 5 (2009): 295–98. The study of primatologists' vocabularies was done by Elizabeth R. Adams and G. W. Burnett, "Scientific Vocabulary Divergence among Female Primatologists Working in East Africa," *Social Studies of Science* 21 (1991): 547–60.

Quotes from German child-rearing manuals are from Alice Miller, *For Your Own Good: Hidden Cruelty in Child-Rearing and the Roots of Violence*, trans. Hildegarde and Hunter Hannum (New York: Farrar Straus Giroux, 1990), 5–57. Figures on corporal punishment come from Murray A. Straus, *Beating the Devil Out of Them: Corporal Punishment in American Families and Its Effects on Children* (New Brunswick, NJ: Transaction, 2001), 206–8. The quote from Alice Miller is found in *For Your Own Good*, 81. Miller takes Hitler's nightmares from a 1940 source: Hermann Rauschning, *The Voice of Destruction: Conversations with Hitler 1940* (New York: G. P. Putnam's Sons, 1940), 256.

For a map of ancestry of Americans, state by state, see Wikipedia's entry "Immigration to the United States," http://en.wikipedia.org/wiki/Immigration_to_the _United_States.

7: ANCESTORS

The chapter epigraph comes from bell hooks, *Sisters of the Yam: Black Women and Self-Recovery* (Cambridge, MA: South End Press, 1993), 175.

In her dissertation Reyda Taylor studies the Indian hobbyists who claim they are "more Indian" than genetically born Indians; her dissertation is "Becoming First Americans" (University of Florida, 2010). There have been huge numbers of discussions of Indian wannabeism and white shamanism over recent decades; perhaps the first was by Cherokee, Quapaw, and Chickasaw writer Geary Hobson, "The Rise of the White Shaman as a New Version of Cultural Imperialism," in *Y'Bird*, ed. Ishmael Reed and Al Young (n.p.: Yardbird, 1977). Rayna Green made the "wannabe" term widespread in "The Tribe Called Wannabe," *Folklore* 99 (1988): 30–55. Cherokee writer Andy Smith criticized New Age practitioners of pseudo-Indian rituals in "For All Those Who Were Indian in a Former Life," *Ms.* 2, no. 3 (November–December 1991): 44–45. The story of trying to patent sweat lodge ceremony comes from L. Whitt, "Indigenous Peoples and the Cultural Politics of Knowledge," in *Issues in Native American Cultural Identity*, ed. Michael Green and Roberta Kevelson (New York: Peter Lang, 1995), 223.

Tobasonakwut's comments were made at the "Language of Spirit" conference sponsored by the SEED Graduate Institute, Albuquerque, New Mexico, August 16, 2010.

The figure of 90 percent deforestation in the mainland United States comes from a University of Michigan curriculum, www.globalchange.umich.edu/global change2/current/lectures/deforest/deforest.html. Quotes about the cutting of forests in the 1790s are found in Alan Taylor, "Wasty Ways," *Environmental History* 3, no. 3 (July 1998): 302–3, and in William Strickland's *Journal of a Tour in the United States of America, 1794-1795* (n.p.: New York Historical Society, 1971), 139. The long quote about driving away original inhabitants and misusing the land only to end up in poverty again comes from William Strickland's *Journal*, 143, 146. The lament about the land being "still as death" comes from Strickland as quoted by Taylor, "Wasty Ways," 303. Taylor's quote on restraint is found in "Wasty Ways," 305.

Judge William Cooper's quote about the "waste of creation" comes from William Cooper and James Fenimore Cooper, *A Guide in the Wilderness: or, The History of the First Settlements in the Western Counties of New York, with Useful Instructions to Future Settlers* (Rochester, NY: G. P. Humphrey, 1879), 7.

The quote from the English Puritan Samuel Mather is found in Samuel Mather, *The Figures or Types of the Old Testament* (London, 1705; reprint, Ann Arbor, MI: University Microfilms, 1966), 159, http://books.google.com. The classic overview of American attitudes toward wilderness is by Roderick Nash, *Wilderness and the American Mind*, 4th ed. (New Haven, CT: Yale University Press, 2001); quote is from 24. Philip Shabecoff traces the opposition between ideas of "wilderness" and "garden" in the first chapter of his excellent history of American environmentalism, *A Fierce Green Fire: The American Environmental Movement*, rev. ed. (Washington: Island Press, 2003).

John Locke's ideas on land as property are found in his *Second Treatise of Civil Government*, 5.32, www.constitution.org/jl/2ndtr05.htm. The Old Germanic idea of working the land is found in Eugene C. Hargrove, *Foundations of Environmental Ethics* (Englewood Cliffs, NJ: Prentice Hall, 1989), 56. Hargrove draws his information on attitudes toward land in old German and English cultures from Denman Waldo Ross, *The Early History of Land-holding among the Germans* (Boston: Soule and Burgbee, 1883), and W. P. Hall, R. G. Albion, and J. B. Pope, *A History of England and the Empire-Commonwealth*, 4th ed. (Lexington, MA: Ginn, 1961).

For discussions of ecology among Greeks and Romans, see the works of J. Donald Hughes: *Pan's Travail: Environmental Problems of the Ancient Greeks and Romans* (Baltimore: Johns Hopkins University Press, 1994) and *An Environmental History of the World: Humankind's Changing Role in the Community of Life*, 2nd ed. (New York: Routledge, 2009). The quote from Cicero comes from *The Nature of the Gods*, trans. C. D. Yonge (New York: Harper & Brothers, 1877), 2:60. The poet

Lucretius's quote about too much land "greedily possessed by mountains and the forests of beasts" is found in Nash, *Wilderness and the American Mind*, 10. The Socrates quote is found in Plato, *Phaedrus*, trans. R. Hackforth, in *The Collected Dialogues of Plato: Including the Letters*, ed. Edith Hamilton and Huntington Cairns (Princeton: Princeton University Press, 1982), 230d. Greek views of nature and ecology are discussed by Robert Sallares in *The Ecology of the Ancient Greek World* (Ithaca, NY: Cornell University Press, 1991). For an unusually engaging look at land usage and politics in the Roman Empire, see Peter Heather, *The Fall of the Roman Empire: A New History of Rome and the Barbarians* (Oxford: Oxford University Press, 2007).

The words of Alexis de Tocqueville on American disregard for wilderness are from his *Journey to America*, trans. George Lawrence, ed. J. P. Mayer (New Haven, 1960), quoted in Nash, *Wilderness and the American Mind*, 23.

Statistics on draining Ohio land can be found in a paper by an emeritus agricultural engineer at Ohio State, Byron H. Nolte, "Ohio Farm Drainage Through the Years—to 2000," available on the web.

Elliott West's words on the pioneer family come from *The Way to the West: Essays on the Central Plains* (Albuquerque: University of New Mexico Press, 1995), 99. Linda Hogan's thoughts on a world "where every place and thing mattered" are found in *The Woman Who Watches Over the World* (New York: Norton, 2001), 63, 30.

Bill McKibben's quip about suburbs came from his interview with Krista Tippett, *On Being*, August 5, 2010, and John O'Donohue's words on place as individuality come from *Anam Cara: A Book of Celtic Wisdom* (New York: HarperCollins, 1997), 85.

Oneida writer Leslie Gray's words come from "The Whole Planet Is the Holy Land," in *Paradigm Wars: Indigenous Peoples' Resistance to Globalization*, ed. Jerry Mander and Victoria Tauli-Corpuz (San Francisco: Sierra Club Books, 2006), 29. Vine Deloria Jr.'s idea of consecrating the land is found in his book *God Is Red: A Native View of Religion*, 2nd ed. (Golden, CO: Fulcrum, 1994), 288. Walter Echo-Hawk's words are found in *A Seat at the Table: Huston Smith in Conversation with Native Americans on Religious Freedom*, ed. Phil Cousineau (Berkeley and Los Angeles: University of California Press, 2006), 34.

Kathleen Dean Moore's luminous essay "Songs in the Night" is found in her *Pine Island Paradox: Making Connections in a Disconnected World* (Minneapolis: Milkweed Editions, 2004). The story of the park ranger is found on 196–97.

The quotes from bell hooks appear in *Sisters of the Yam*, 190, 175.

Linda Hogan's words on the values taught by Native ancestors come from *Woman Who Watches over the World*, 30.

8: PERALTA CREEK

Shunryu Suzuki's words are found in *Zen Mind, Beginner's Mind* (New York: Weatherhill, 1970), 29.

The Rarámuri way of seeing nature as kinfolk joined in a great circle of breath is explained by Enrique Salmón, "Kincentric Ecology: Indigenous Perceptions of the Human-Nature Relationship," *Ecological Applications* 10, no. 5 (October 2000): 1327–32. Quotes are found on 1328 and 1331.

Photos of the first Butters Canyon cleanup day are still online at www.bcconservancy.org/canyon_events/canyon_cleanup/Clean2000Photos/index.htm.

Michael Pollan spoke on Twinkies and rot during a book tour stop in Boulder, CO, May 2009. For a fuller report, see my blog post "In Defense of Food," http://priscillastuckey.com/2009/05/23/in-defense-of-food/.

A brief introduction to fracking and the destruction of water appears on the site of the nonprofit Earthworks, www.earthworksaction.org/issues/detail/hydraulic_fracturing_101.

For the beginning of the recent wilderness debate, see William Cronon, "The Trouble with Wilderness; or, Getting Back to the Wrong Nature," in *Uncommon Ground: Rethinking the Human Place in Nature*, ed. William Cronon (New York: Norton, 1995), 69–90. This essay also inaugurated the new journal *Environmental History*, and its first issue, in 1996, is full of responses, rebuttals, and conversations about Cronon's piece. A retrospective of the wilderness debate can be found in J. Baird Callicott and Michael P. Nelson, eds., *The Great New Wilderness Debate* (Athens: University of Georgia Press, 1998).

For more information on the land trust movement, see Richard Brewer, *Conservancy: The Land Trust Movement in America* (Lebanon, NH: University Press of New England, 2003).

The quote from N. Scott Momaday comes from "A First American Views His Land," in *At Home on the Earth: Becoming Native to Our Place*, ed. David Landis Barnhill (Berkeley and Los Angeles: University of California Press, 1999), 28.

The Butters Canyon Conservancy (originally called the Butters Land Trust) maintains a website at www.bcconservancy.org. News stories and videos about its accomplishments include these: Carolyn Jones, "Neighbors' Work to Save Butters Canyon Pays Off," *San Francisco Chronicle*, March 5, 2010, www.sfgate.com/cgi-bin/article.cgi?f=/c/a/2010/03/05/BAV51CAVA1.DTL; and Allie Rasmus, KTVU video story, March 14/15, 2010, www.ktvu.com/search/?q=Allie+Rasmus%2C+Butters+Canyon.

9: SHVANA

"The Mountains, I Become Part of It," is quoted by Joseph Epes Brown, "Becoming Part of It," in *I Become Part of It: Sacred Dimensions in Native American Life*, ed. D. M. Dooling and Paul Jordan-Smith (San Francisco: HarperSanFrancisco, 1989), 20.

"Enlightenment—nothing special" comes from Shunryu Suzuki, *Zen Mind, Beginner's Mind*, ed. Trudy Dixon (New York and Tokyo: Weatherhill, 1970), 47–48.

Malidoma Somé's recounting of Dagara cosmology and attitudes toward language is found in *The Healing Wisdom of Africa* (New York: Jeremy P. Tarcher, 1999), 50.

My telling of the story of Nachiketa and Death is based on three translations: Swami Gambhirananda, trans., *Katha Upanisad* (Calcutta: Advaita Ashrama, 1980); Swami Sarvananda, trans., *Kathopanisad* (Madras: Sri Ramakrishna Math, 1975); and Eknath Easwaran's lovely translation in *The Upanishads*, 2nd ed. (Tomales, CA: Nilgiri Press, 2007). All quotes that are set as poetry lines are Easwaran's translation. "Ask for sons and grandsons" is from Katha Upanishad 1.1.23 (pp. 73–74); "Sharp like a razor's edge" is from Katha Upanishad 1.3.14 (p. 82); "When all the knots" is from the Brihadaranyaka Upanishad 4.4.6 (p. 115). Other quotes from the Upanishads are my own paraphrases based on the three translations.

Mark Bittner tells the story of Tupelo in the documentary *The Wild Parrots of Telegraph Hill*, film by Judy Irving (2005), and also on his blog, "The Difficulties of the Zen Waterfall," Views from a Hill, http://markbittner.wordpress.com/2011/04/25/the-difficulties-of-the-zen-waterfall.

For a beautiful depiction of the healing in both body and mind that can take place through the ritual of the sand painting, see Leslie Marmon Silko's novel *Ceremony* (New York: Viking Press, 1977). Paula Gunn Allen's words on the identity of people and land are found in "IYANI: It Goes This Way," in *The Remembered Earth: An Anthology of Contemporary Native American Literature*, ed. Geary Hobson (Albuquerque: University of New Mexico Press, 1979), 191.

Statistics on embalming and burial come from Rachel Saslow, "Green Burials Are Gaining Traction in the Washington Area," *Washington Post*, June 6, 2011, www.washingtonpost.com/national/environment/green-burials-are-gaining-traction-in-the-washington-area/2011/04/26/AGLfhZKH_story.html.

Donna Haraway's thoughts on death as necessary to life are found in Donna Haraway, interview by Thyrza Goodeve, *How Like a Leaf: An Interview with Donna Haraway* (New York: Routledge, 2000), 115–16.

John Muir's thoughts on a "stingless" death are from *A Thousand-Mile Walk to the Gulf*, ed. William Frederic Badè (Boston: Houghton Mifflin, 1916), 70–71. That book was edited by Muir and Badè shortly before Muir's death and was published posthumously. The original draft from Muir's journal as a thirty-year-old is quoted and discussed in Steven J. Holmes, *The Young John Muir: An Environmental Biography* (Madison: University of Wisconsin Press, 1999), 185.

Ancient Christian and Jewish views of resurrection come from Kevin J. Madigan and Jon D. Levenson, *Resurrection: The Power of God for Christians and Jews* (New Haven, CT: Yale University Press, 2008). Notions of personal immortality among Roman Christians and pagans are found in John Matthews, "Four Funerals and a Wedding," in *Transformations of Antiquity: Essays for Peter Brown*, ed. Philip Rousseau and Manolis Papoutsakis (Burlington, VT: Ashgate Publishing, 2009), 129–46.

The Christian ascetics are often overlooked as the authors of the first nature writing and forerunners of the environmental movement. A scholarly essay that gives them their due is Judith Adler's "Cultivating Wilderness: Environmentalism and Legacies of Early Christian Asceticism," *Comparative Studies in Society and History* 48 (2006): 4–37. Roger Sorrell recaps this history briefly in *St. Francis of Assisi and Nature: Tradition and Innovation in Western Christian Attitudes Toward the Environment* (New York: Oxford University Press, 1988), 19. Quotes from Basil the Great are found in Sorrell, 19. For the history of Christian attitudes toward wilderness, see George H. Williams, *Wilderness and Paradise in Christian Thought* (New York: Harper & Brothers, 1962); Williams was the basis for much of Roderick Nash's first chapter in *Wilderness and the American Mind*, 4th ed. (New Haven, CT: Yale University Press, 2001). The classic study of Western attitudes toward nature is Clarence J. Glacken, *Traces on the Rhodian Shore: Nature and Culture in Western Thought from Ancient Times to the End of the Eighteenth Century* (Berkeley and Los Angeles: University of California Press, 1967). Glacken was one of the first to draw connections between the ascetics and environmentalism.

Details of the death of St. Francis were written down by people close to him within a few years of his death and are discussed by Adrian House, *Francis of Assisi: A Revolutionary Life* (Mahwah, NJ: Paulist Press, 2000), 277–79; and Edward A. Armstrong, *St. Francis, Nature Mystic: The Derivation and Significance of the Nature Stories in the Franciscan Legend* (Berkeley and Los Angeles: University of California Press, 1973), 96, 239.

10: FAMILY

The chapter epigraph is from Sharon Butala's memoir of learning to know life on the Canadian prairie, *Perfection of the Morning: A Woman's Awakening in Nature* (St. Paul, MN: Hungry Mind Press, 1994), 192.

The Sweet Honey in the Rock lyric comes from "I Be Your Water," in the album *In This Land* (Earthbeat, 1992).

The book I was copyediting at the time I traveled to DC to visit Tim was Dan Millman, *The Journeys of Socrates: An Adventure* (San Francisco: HarperSanFrancisco, 2005); the quote is found on 252.

The Terry Tempest Williams quote is from her essay "Winter Solstice at the Moab Slough," in *An Unspoken Hunger: Stories from the Field* (New York: Pantheon, 1994), 64.

The quote from Pema Chödrön comes from *When Things Fall Apart: Heart Advice for Difficult Times* (Boston: Shambhala, 1997), 10.

Rumi's poem "The Guest House" is translated by Coleman Barks, *The Essential Rumi* (San Francisco: HarperSanFrancisco, 1995), 109.

The complete text of Inanna's trip to the underworld can be found online: J. A. Black, G. Cunningham, J. Ebeling, E. Flückiger-Hawker, E. Robson, J. Taylor, and G. Zólyomi, eds., *The Electronic Text Corpus of Sumerian Literature* (Oxford: Oxford University Press, 1998–2006), http://etcsl.orinst.ox.ac.uk. The translation of the opening line of the story, "She opened her ear to the great below," comes from D. Wolkstein and Samuel Noah Kramer, *Inanna, Queen of Heaven and Earth: Her Stories and Hymns from Sumer* (New York: Harper & Row, 1983), 52.

11: BRIO

The chapter epigraph comes from Sheri Hostetler's poem "Instructions," in the anthology *A Cappella: Mennonite Voices in Poetry*, ed. Ann Hostetler (Iowa City: University of Iowa Press, 2003), 126.

Roderick Nash's etymology of the word *wilderness* is found in *Wilderness and the American Mind*, 4th ed. (New Haven, CT: Yale University Press, 2001), 1.

David Mas Masumoto's comments on control in American life come from *Epitaph for a Peach: Four Seasons on My Family Farm* (San Francisco: HarperSanFrancisco, 1996), 64.

Good information on the canals and dikes of early Mesopotamia can be found in D. T. Potts, *Mesopotamian Civilization: The Material Foundations* (Ithaca,

NY: Cornell University Press, 1997), chap. 1. The text of the *Enuma elish* is from James Pritchard, ed., *The Ancient Near East: An Anthology of Texts and Pictures* (Princeton: Princeton University Press, 2011), 28–35.

Walter Brueggemann is a Christian scholar of Hebrew scriptures who reviews the emerging consensus among biblical scholars that the theology of the Old Testament was shaped above all by the exile in Babylon; see his *Theology of the Old Testament: Testimony, Dispute, Advocacy* (Minneapolis: Augsburg Fortress, 1997), esp. 74–75.

Physicist Victor Mansfield's discussion of determinism and causality comes from *Synchronicity, Science, and Soul-Making: Understanding Jungian Synchronicity Through Physics, Buddhism, and Philosophy* (Chicago: Open Court Press, 1995), 73. I am grateful to Mansfield for suggesting that free will connects with quantum mechanics. He explores the similarities and differences between the notions of causality in quantum mechanics and Middle Way Buddhism in *Tibetan Buddhism and Modern Physics: Toward a Union of Love and Knowledge* (West Conshohocken, PA: Templeton Foundation Press, 2008). Laplace's misplaced confidence in absolute causality can be found in Pierre-Simon Laplace, *A Philosophical Essay on Probabilities*, trans. from the 5th French edition by Andrew Dale (New York: Springer-Verlag, 1995), 2. Bohr's retort to Einstein about God and dice is reported by John Wheeler, "Law Without Law," in *Quantum Theory and Measurement*, ed. John Wheeler and Wojciech Zurek (Princeton: Princeton University Press, 1983), 188. The quote from Mansfield is from *Synchronicity*, 78.

The translation of Calvin I used is John Calvin, *Institutes of the Christian Religion*, trans. John T. McNeill and Ford Lewis Battles (Louisville, KY: John Knox Press, 1960), 3.23.2, 2:949. I changed the final word of the quote, *righteous*, to *just*.

Mary Oliver's poem "Shadows" is found in *Dream Work* (New York: Atlantic Monthly Press, 1986), 17.

On the links between colonization and the promulgators of the scientific revolution, one example is Robert Boyle, a member of the Royal Society. It is well known that he inherited his wealth from a father who had amassed it by helping the English crown confiscate and colonize Irish lands in the 1500s. Likewise, the wealthy John Locke, who wrote the template for English property law, made a profit from investing in the Royal Africa Company, which engaged in slave trade on behalf of England, and he also owned shares in the slave-trading Bahama Adventurers.

Quotes from Francis Bacon are found in the *Sylva Sylvarum*, from *The Works of Francis Bacon*, ed. James Spedding, Robert Leslie Ellis, and Douglas Denon Heath (London, 1627; reprint, London, 1857), 2.10.900.

Assurbanipal's report of his early education is quoted in Eckart Frahm, "Royal Hermeneutics: Observations on the Commentaries from Assurbanipal's Libraries at

Ninevah," *Papers of the 49th Rencontre Assyriologique Internationale*, pt. 1 (2004): 45. An overview of Mesopotamian divination can be found in Frederick H. Cryer, *Divination in Ancient Israel and Its Near Eastern Environment: A Socio-historical Investigation* (Sheffield, England: JSOT Press, 1994). The term *omen science* comes from Francesca Rochberg, "Conditionals, Inference, and Possibility in Ancient Mesopotamian Science," *Science in Context* 22, no. 1 (2009): 5–25.

On a holistic universe and its relation to omens and magic, see Wim van Binsbergen and Frans Wiggermann, "Magic in History: A Theoretical Perspective, and Its Application to Ancient Mesopotamia," in *Mesopotamian Magic: Textual, Historical, and Interpretative Perspectives*, ed. Tzvi Abusch and Karel van der Toorn (Gröningen: STYX Publications, 1999), 1–35.

On the social structure of Assyrian society, see Marc Van de Mieroop, *A History of the Ancient Near East ca. 3000–323 BC*, 2nd ed. (Malden, MA: Blackwell, 2007), 236. Assyrian omens about cats come from Ann K. Guinan, "Divination," in *The Context of Scripture: Canonical Compositions from the Biblical World*, ed. William W. Hallo (Leiden, the Netherlands: E. J. Brill, 2003), 1:424. For omens in the Roman Empire, see Georg Luck, trans., *Arcana Mundi: Magic and the Occult in the Greek and Roman Worlds; A Collection of Ancient Texts*, 2nd ed. (Baltimore: Johns Hopkins University Press, 2006). The big collection of magical spells and rituals from antiquity is Hans Dieter Betz, ed., *The Greek Magical Papyri in Translation, including the Demotic Spells*, 2nd ed. (Chicago: University of Chicago Press, 1992).

The most exhaustive compendium of cultural beliefs about cats is Carl van Vechten's *The Tiger in the House: A Cultural History of the Cat* (New York: New York Review of Books, 1920). The charming story of the Penitent's cat ascending to heaven is told on 180–81. The story comes from the French writer Moncrif and so may be a story about France more than a story about India. For more scientific views of cats, see the fascinating book *The Domestic Cat: The Biology of Its Behavior*, ed. Dennis C. Turner and Patrick Bateson, 2nd ed. (Cambridge: Cambridge University Press, 2000).

Anthropologist David L. Browman calls attention to the connection between Andean omens and Western scientific observations in "Central Andean Views of Nature and the Environment," in *Nature Across Cultures: Views of Nature and the Environment in Non-Western Cultures*, ed. Helaine Selin (Dordrecht, the Netherlands: Kluwer Academic Publishers, 2003), 290–92. The group of anthropologists and meteorologists who studied Andean stargazing as compared to satellite observations published their findings in *Nature* magazine: see Benjamin S. Orlove, John C. H. Chiang, and Mark A. Cane, "Forecasting Andean Rainfall and Crop Yield from the Influences of El Niño on Pleiades Visibility," *Nature* 403 (2000):

68–71. The *Nature* report is available from the website of Benjamin Orlove, www .des.ucdavis.edu/faculty/orlove/publications/recent1.html.

Many alternative measures of national health are being proposed to replace the GDP, such as the Genuine Progress Indicator (GPI), now instituted by law in Vermont, the Gross National Happiness (GNH) index, and the Happy Planet Index. Information on these and other measures is available on the web. My thanks to James Pittman and Timothy Falb for helping me think through the relationship between economics and superstition.

The equation describing predator-prey relationships is the Lotka-Volterra equation, and it has been applied by researchers at the National Oceanic and Atmospheric Administration (NOAA, Boulder, CO) and the Weizmann Institute of Science (Israel) to fluctuating rain and cloud patterns. See the Weizmann Institute of Science, "Eat, Prey, Rain: New Model of Dynamics of Clouds and Rain Is Based on a Predator-Prey Population Model," *ScienceDaily*, July 25, 2011, www.science daily.com/releases/2011/07/110725091726.htm; and Linda Joy, "NOAA Scientists Uncover Oscillating Patterns in Clouds," *NOAA Research*, August 11, 2011, www .oar.noaa.gov/news/archive/2010/oscillating_patterns_clouds.html. The original research article is I. Koren and G. Feingold, "Aerosol-Cloud-Precipitation System as a Predator-Prey Problem," *Proceedings of the National Academy of Sciences* (2011).

On the self-organizing abilities of living organisms and nonliving systems, see Scott Camazine, "Patterns in Nature," *Natural History* (June 2003), 34–41, also his book: Scott Camazine, Jean-Louis Deneubourg, Nigel R. Franks, James Sneyd, Guy Theraula, and Eric Bonabeau, *Self-Organization in Biological Systems* (Princeton: Princeton University Press, 2002). The theoretical biologist best known for his books on spontaneous order and self-organization is Stuart A. Kauffman; his titles include *The Origins of Order: Self-Organization and Selection in Nature* (Oxford: Oxford University Press, 1993); *At Home in the Universe: The Search for Laws of Self-Organization and Complexity* (Oxford: Oxford University Press, 1995); and *Reinventing the Sacred: A New View of Science, Reason, and Religion* (New York: Basic Books, 2008). The Kauffman quote comes from *Reinventing the Sacred*, xi. Nobel Prize–winning physicist Robert Laughlin takes up the theme of self-organization in *A Different Universe: Reinventing Physics from the Bottom Down* (New York: Basic Books, 2005).

My discussion of quantum mechanics is shaped by these sources: P. C. W. Davies and J. R. Brown, *The Ghost in the Atom: A Discussion of the Mysteries of Quantum Physics* (Cambridge: Cambridge University Press, 1986), quote from 28; John Wheeler, "Law Without Law," in *Quantum Theory and Measurement*, ed. John Wheeler and Wojciech Zurek (Princeton: Princeton University Press, 1983),

182–214; Arthur Zajonc, *Catching the Light: The Entwined History of Light and Mind* (New York: Oxford University Press, 1993); and most of all Karen Barad, *Meeting the Universe Halfway: Quantum Physics and the Entanglement of Matter and Meaning* (Durham, NC: Duke University Press, 2007). The image of the dark sidewalk and streetlamps comes from Victor Mansfield, who said it was a man, not a cat, appearing under the lights; see *Synchronicity, Science, and Soul-Making*, 97.

The quote about the sacred hoop, with everything having a place in the circle, comes from Paula Gunn Allen's mother, in Allen, *The Sacred Hoop: Recovering the Feminine in American Indian Traditions*, 2nd ed. (Boston: Beacon Press, 1992). Wheeler's quote about the community of observer-participants comes from "Law Without Law," 202.

Susan Griffin's fine essay "A Collaborative Intelligence" appears in *The Eros of Everyday Life: Essays on Ecology, Gender and Society* (New York: Doubleday, 1995); quotes are from 39.

12: EARTH

The chapter epigraph, Aldo Leopold's famous quote about land as a community, comes from the Foreword to *A Sand County Almanac* (Oxford: Oxford University Press, 1949), viii.

Most of my information on Aboriginal culture and land management practices comes from Deborah Bird Rose: *Nourishing Terrains: Australian Aboriginal Views of Landscape and Wilderness* (Canberra: Australian Heritage Commission, 1996) and *Dingo Makes Us Human: Life and Land in an Australian Aboriginal Culture* (Cambridge: Cambridge University Press, 1992). Rose's words that "country is home, and peace; . . . heart's ease" comes from *Nourishing Terrains*, 1. The quote about big fires comes from *Nourishing Terrains*, 66. Daly Pulkara's view of wild land is found in *Nourishing Terrains*, 19. April Bright's comment on looking after country is from *Nourishing Terrains*, 49. Doug Campbell's comments on the Aboriginal Law residing in the ground comes from *Dingo*, 56.

On the various theories of human separation from land: the Paul Shepard quote is from *Coming Home to the Pleistocene* (Washington, DC: Island Press, 1998), 81. Shepard's *Nature and Madness* (San Francisco: Sierra Club Books, 1982) is built on the premise of social decline beginning with agriculture. The theory that cities spelled doom for connections to nature is found in J. Donald Hughes, *An Environmental History of the World: Humankind's Changing Role in the Community of Life*, 2nd ed. (London: Routledge, 2009), chap. 3. The term *urban revolution* comes from the famous archeologist of prehistory, V. Gordon Childe. Ecological philosopher David Abram details the changes in perception caused by literacy in *The Spell of the Sensuous: Perception and Language in a More-Than-Human World* (New York: Random House, 1996); see also Abram, *Becoming Animal: An Earthly Cosmology* (New York: Random House, 2010). Abram draws in part on Eric Havelock and Walter Ong, both of whom explored the shift in cognition that took place in ancient Greece during the transition to literacy, especially the erosion of relationships among humans. Havelock's influential book is *The Muse Learns to Write: Reflections on Orality and Literacy from Antiquity to the Present* (New Haven, CT: Yale University Press, 1986), especially chap. 10, and Walter Ong's is *Orality and Literacy: The Technologizing of the Word* (London: Methuen & Co., 1982), esp. 103–4.

The study of income inequality among five contemporary hunter-gatherer groups is Eric Alden Smith, Kim Hill, Frank W. Marlowe, David Nolin, Polly Wiessner, Michael Gurven, Samuel Bowles, Monique Borgerhoff Mulder, Tom Hertz, and Adrian Bell, "Wealth Transmission and Inequality Among Hunter-Gatherers," *Current Anthropology* 51, no. 1 (February 2010): 19–34, available at NIH Public Access, www.ncbi.nlm.nih.gov/pmc/articles/PMC2999363/?tool=pubmed. The idea that Westerners can't take a joke about their property comes from Deborah Bird Rose in *Dingo Makes Us Human,* 167.

On practicing give-and-take: anthropologist Tim Ingold discusses Aboriginal landownership as obligating one to share; see Tim Ingold, *The Appropriation of Nature: Essays on Human Ecology and Social Relations* (Iowa City: University of Iowa Press, 1987); quote is from 134. For the possibilities of redefining property ownership as trusteeship, see Polly Higgins, *Eradicating Ecocide: Exposing the Corporate and Political Practices Destroying the Planet and Proposing the Laws Needed to Eradicate Ecocide* (London: Shepheard-Walwyn, 2010).

Joshua Farley presented "Towards a Just Distribution of Resources" at the Gund Institute for Ecological Economics, University of Vermont, February 2, 2003, video available at www.postcarbon.org/video/371487. Farley discusses his grad school days in Lissa Harris, "Rewriting the Book on Economics," *Grist,* April 9, 2003, www.grist.org/article/the18.

Information on the potlatch comes from Ronald Trosper, "Incentive Systems That Support Sustainability: A First Nations Example," *Conservation Ecology* 2, no. 2 (1998): 11, available at Ecology and Society, www.ecologyandsociety.org/vol2/iss2/art11. He expanded the article into a book: *Resilience, Reciprocity, and Ecological Economics: Northwest Coast Sustainability* (London: Routledge, 2009). Trosper uses the reciprocal system of the Northwest Coast people to answer the charge of some economists that reciprocal systems are utopian; see Trosper, *Resilience*, 21–22, discussing J. Elster and K. O. Moene, eds., *Alternatives to Capitalism* (Cambridge: Cambridge University Press, 1989), 1–35.

On the Year of Jubilee among Israelites and the clean-slate laws in Mesopotamia, see Michael Hudson, "It Shall Be a Jubilee unto You," *Yes!* magazine, September 30, 2002, www.yesmagazine.org/23livingeconomy/hudson.htm. Hudson's scholarly essay on the Jubilee is "'Proclaim Liberty Throughout the Land': The Economic Roots of the Jubilee," *Bible Review* 15, no. 1 (February 1999): 26–33, 44. On land tenure in Palestine at the time of Jesus, see Hudson, "Proclaim Liberty," as well as Herman C. Waetjen, "Intimations of the Year of Jubilee in the Parables of the Wicked Tenants and Workers in the Vineyard," *Christian Century*, May 20–27, 1998, 524–31.

On privatization as contributing to the economic collapse of the Roman Empire, see Michael Hudson, "Entrepreneurs: From the Near Eastern Takeoff to the Roman Collapse," in *The Invention of Enterprise: Entrepreneurship from Ancient Mesopotamia to Modern Times*, ed. David S. Landes, Joel Mokyr, and William J. Baumol (Princeton: Princeton University Press, 2010), 8–39, available on the web as "History of Debt and Property from the Ancient East," www.prosper.org .au/2010/07/20/history-of-debt-and-property-from-the-ancient-east/. Historians of ancient Rome now agree that there was a great deal of continuity in social and economic arrangements before and after the fall of the empire. Peter Heather argues that though the central bureaucracy changed from Roman to Germanic rule, the rural landowners escaped with the bulk of their estates intact. See Peter Heather, *The Fall of the Roman Empire: A New History of Rome and the Barbarians* (Oxford: Oxford University Press, 2007). Heather provides the figure of less than 5 percent of the people making up the landowning class in the fourth century (133, 138).

Translations of Ambrose's Latin sermon "On Naboth" are my own, guided by the translation and commentary of Martin R. P. McGuire, *S. Ambrosii De Nabuthae* (Washington, DC: Catholic University of America, 1927). Quotes are from paragraphs 2.1; 3.11; 2.4; and 14.58–59. Ambrose was not so liberal minded in other ways. In the same sermon he made sexist generalizations to the effect that all wives were greedy for expensive jewelry: "Women even delight in shackles, as long as they are fastened with gold" (6.26). His other writings are full of diatribes against Jews and pagans. Details of Ambrose's consecration as bishop come from Neil B. McLynn, *Ambrose of Milan: Church and Court in a Christian Capital* (Berkeley and Los Angeles: University of California Press, 1994). Ambrose's attitudes toward rich and poor are central in the work of Vincent R. Vasey, S.M., *The Social Ideas in the Works of St. Ambrose: A Study on De Nabuthe* (Rome: Institutum Patristicum

"Augustinianum," 1982). Peter Heather sheds light on Ambrose's education in rhetoric in *Fall of the Roman Empire*, chaps. 1, 3.

Aldo Leopold's question "Do economists know about lupine?" comes from *Sand County Almanac*, 102.

Rumi's couplet about "the speech of water" (*Masnavi* 1.3292) was translated by Richard Foltz and Frederick Denny, to whom I am grateful. Rumi's couplet about rust on the mirror is quoted in Mohammed Rustom, "The Metaphysics of the Heart in the Sufi Doctrine of Rumi," *Studies in Religion/Sciences Religieuses* 37, no. 3 (2008): 4.

Winona LaDuke's thoughts on sharing resources come from *A Seat at the Table: Huston Smith in Conversation with Native Americans on Religious Freedom*, ed. Phil Cousineau (Berkeley and Los Angeles: University of California Press, 2006), 48–49.

EPILOGUE: NIGHT

The complete text of George Eliot's *Middlemarch* is available online at http://eliot.thefreelibrary.com/Middlemarch. The epigraph appears in chap. 20.

PERMISSIONS

INDEX